ENVIRONMENT

A plea for a cleaner, greener planet, this book will appeal to readers across generations and geographical boundaries.

R. Rajagopalan has taught for over thirty years at IIT Kanpur and IIT Madras. He has written and edited several books, including twelve textbooks on environmental education for schools and colleges. Two of his children's books have won national awards. He lectures and conducts workshops on environmental issues all over India.

ENVIRONMENT

R. Rajagopalan

OXFORD
UNIVERSITY PRESS

OXFORD
UNIVERSITY PRESS

Oxford University Press is a department of the University of Oxford.
It furthers the University's objective of excellence in research, scholarship,
and education by publishing worldwide. Oxford is a registered trademark of
Oxford University Press in the UK and in certain other countries

Published in India by
Oxford University Press
YMCA Library Building, 1, Jai Singh Road, New Delhi 110001, India

First published in 2011

ISBN-13: 978-0-19-807699-5
ISBN-10: 0-19-807699-1

Typeset in ITC Legacy Serif Std 10/13
by Eleven Arts, Keshav Puram, Delhi 110 035
Printed in India by G.H. Prints Pvt Ltd., New Delhi 110020

Illustrations by Kallol Majumder

To
the innumerable and inspiring
heroes of the environment,
conserving nature in
villages and local communities
all over the world

Dang Thi Hai

Dear Hai,

It is never too late to
find the true purpose of
one's life.

Best wishes

R. [illegible]

Feb 2013

CONTENTS

THE REMEDY

PREFACE

We are truly living in extraordinary times. Our actions of today will have an impact (mostly negative) on the planet for decades to come and we will pay a price for them. If you are under forty, you will see in your lifetime a vastly different world–and not a comfortable one at that. This book tells you a part of that story.

If today is a typical day on planet Earth, we will lose 116 sq. miles of rainforest, or about an acre a second. We will lose another 72 sq. miles to encroaching deserts as a result of human mismanagement and overpopulation. We will lose 40 to 100 species, and no one knows whether the number is 40 or 100. Today the human population will increase by 2,50,000. And today we will add 2,700 tons of chlorofluorocarbons to the atmosphere and 15 million tons of carbon. Tonight the Earth will be a little hotter, its waters more acidic, and the fabric of life more threadbare.

The truth is that many things on which your future health and prosperity depend are in dire jeopardy: climate stability, the resilience and productivity of natural systems, the beauty of the natural world, and biological diversity.

That was a statement made in 1990 by the American environmentalist and educator, David Orr. Two decades later, the indicators have only got much worse.

That the planet is in peril is known to thousands of scientists, academics, authors, activists, NGOs, school children, poor farmers, indigenous people, and other groups. It is also known to many corporate leaders, administrators, and even politicians. United Nations agencies and international NGOs (such as Greenpeace and World Wide Fund for Nature) release reports almost daily on the scale of environmental destruction, make dire predictions, and call for immediate action. We often read about 'crossing tipping points', 'impending climate catastrophe', and 'small windows of opportunity to reverse the downslide'.

Yet, what are we doing today?

- Governments, politicians, and most economists swear by the wisdom of unending economic growth that is fast depleting our natural resources.
- The citizens of the industrialized North continue their mindless consumption of goods and services.
- The rest of us (especially the Indians and the Chinese) are desperately trying to catch up with the consumption levels of the US and others.
- Together we generate millions of tons of waste and dump them on land, waterways, and seas.
- Our carbon dioxide emissions have crossed safe limits and continue to soar.
- Corporations resist all environmental regulations.
- Governments, keeping elections in mind, are unwilling to introduce tough measures to conserve the environment.
- In international negotiations on climate change, the South blames the North, the North points to the South, and nobody accepts any emission cuts.

It is not just the environmental crisis, there are other danger signs too. Global financial meltdown, collapsing economies, failed states, runaway military expenditures, increasing inequalities, soaring food prices, large scale corruption, rising public anger, and protest movements—the list goes on. Ultimately, however, all these troubles and the ecological decline are interconnected. But that is another story.

Why is there so little action to save the environment? Have YOU done anything about it? If you have not, I cannot blame you. It took me many years to realize the gravity of the situation, though I belong to the very generation that has plundered the planet.

I can assure you, however, that once you begin to understand what is going on, you will be a changed person. You cannot then get ecology and environment out of your mind. It happened to me. I have also seen it happen to those who read inspiring books on the subject, to those who attended workshops in Auroville and elsewhere.

This book addresses questions such as:

- What is state of the global and Indian environment? What is the scorecard with respect to water, waste, energy, biological diversity, population, and food?
- What is likely to happen if we continue 'business as usual'?
- How did we end up in this crisis? How far is humanity responsible for it?
- What has the world done about the crisis so far?

- What can you do to set things right?
- Wherein lies hope for the future?

The book is aimed at senior school students and adults of any age. It includes many true stories from India and abroad. The prologues contain inspiring messages. There are also poems, questions with surprising answers, and end pieces to set you thinking. Publications, websites, and films are listed at the end of each chapter. Since environment can be a depressing topic, there is also some humour through quotations and cartoons.

I do suggest a number of action items for you, such as 'Change to CFL' and 'Harvest rainwater', but it is high time now for bigger efforts. I end the book on a hopeful note with inspiring stories and a call to change your mindset.

In 1967, Martin Luther King, Jr. said in the context of the Vietnam War and the Civil Rights Movement:

We are now faced with the fact that tomorrow is today. We are confronted with the fierce urgency of now. In this unfolding conundrum of life and history there is such a thing as being too late. Procrastination is still the thief of time. Life often leaves us standing bare, naked and dejected with a lost opportunity .... Over the bleached bones and jumbled residue of numerous civilizations are written the pathetic words: 'Too late'.

However, every word of his statement is valid today in the environmental context. Let me end with another quotation, one from the environmentalist Paul Hawken:

Nature beckons you to be on her side. You couldn't ask for a better boss. The most unrealistic person in the world is the cynic, not the dreamer. Hope makes sense only when it doesn't make sense to be hopeful. This is your century. Take it and run as if your life depends on it.

Bangalore
July 2011

R. RAJAGOPALAN

P.S. I invite you to write to me at rrgopalan2005@gmail.com. You can also access my blog at www.profrr.blogspot.com.

INTRODUCTION

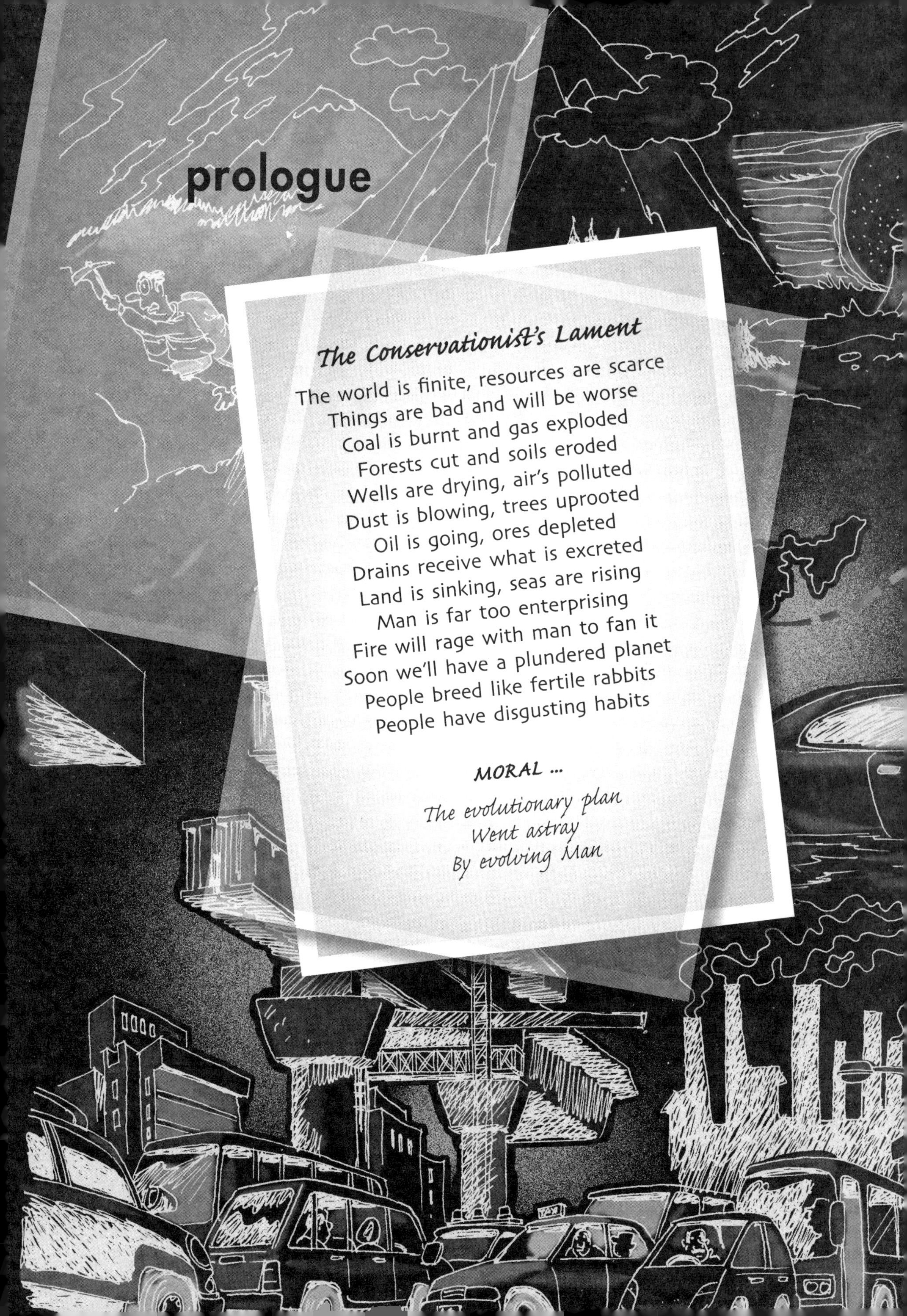
prologue
The Conservationist's Lament
The world is finite, resources are scarce
Things are bad and will be worse
Coal is burnt and gas exploded
Forests cut and soils eroded
Wells are drying, air's polluted
Dust is blowing, trees uprooted
Oil is going, ores depleted
Drains receive what is excreted
Land is sinking, seas are rising
Man is far too enterprising
Fire will rage with man to fan it
Soon we'll have a plundered planet
People breed like fertile rabbits
People have disgusting habits
MORAL ...
The evolutionary plan
Went astray
By evolving Man

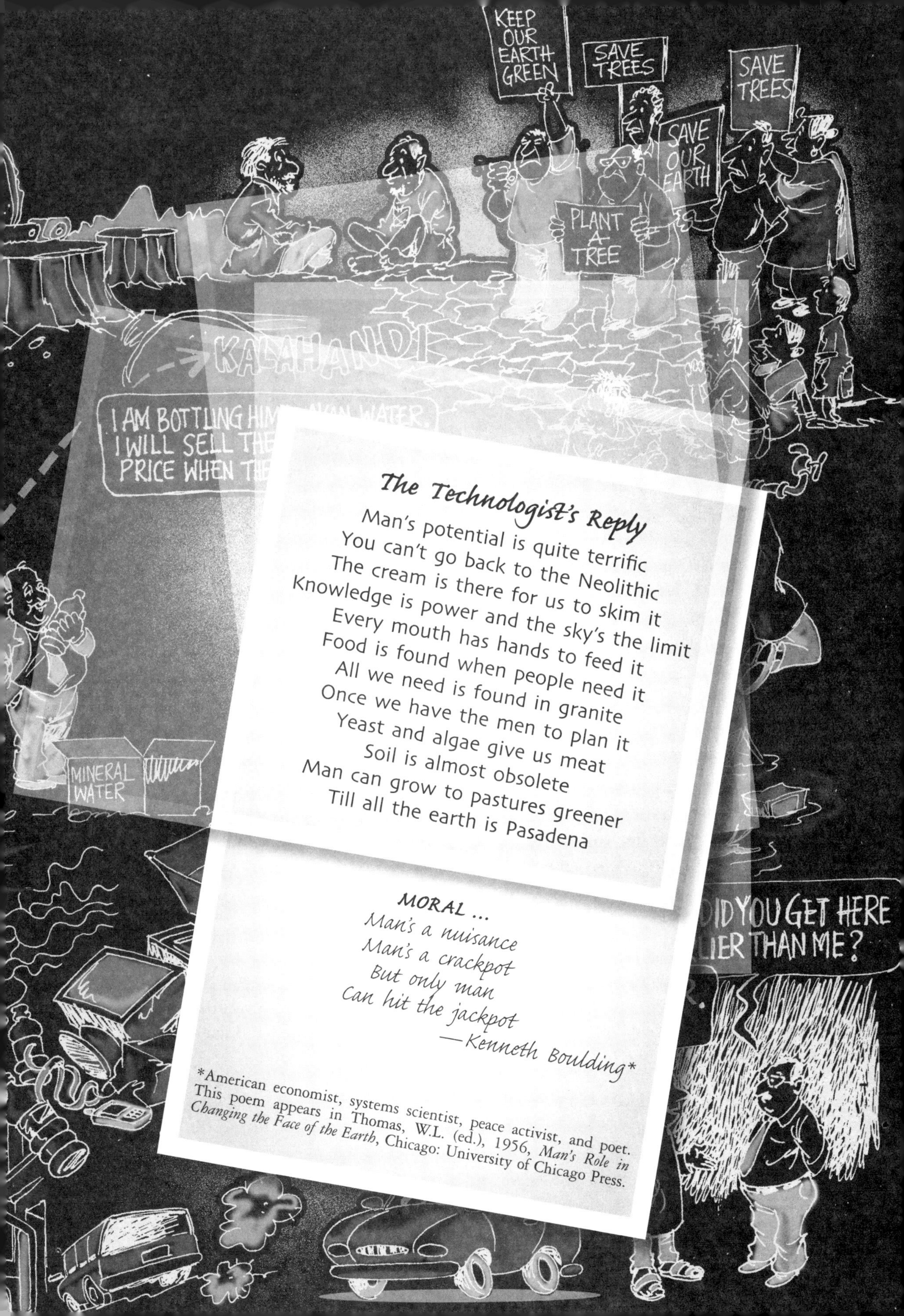

The Technologist's Reply

Man's potential is quite terrific
You can't go back to the Neolithic
The cream is there for us to skim it
Knowledge is power and the sky's the limit
Every mouth has hands to feed it
Food is found when people need it
All we need is found in granite
Once we have the men to plan it
Yeast and algae give us meat
Soil is almost obsolete
Man can grow to pastures greener
Till all the earth is Pasadena

MORAL ...

Man's a nuisance
Man's a crackpot
But only man
Can hit the jackpot

—Kenneth Boulding*

*American economist, systems scientist, peace activist, and poet. This poem appears in Thomas, W.L. (ed.), 1956, *Man's Role in Changing the Face of the Earth*, Chicago: University of Chicago Press.

1 PLANET IN PERIL
Three Stories

If it's just the noise of progress that is beating in our ears
We could look beyond the turbulence and soothe our gnawing fears.
Man is drowning in his own success, and hapless is his hope
If our science and technology is but a rotten rope.

—KENNETH E. BOULDING, 'A BALLAD OF ECOLOGICAL AWARENESS'

THE STATE OF AN ECOSYSTEM: PARADISE THEN, PLUNDERED NOW

'If there is paradise anywhere on earth, it is here, it is here, it is here.' So said the Mughal Emperor Jehangir about Kashmir in the Himalayas. If, however, he were to return to the mountain range now, he would be shocked and saddened at its condition.

The majestic Himalayas are not merely a range of mountains. They are the cradle of Indian civilization and have exerted great influence over Indian thought, culture, and, of course, the environment.

If the Himalayas were not there, the rain clouds from the Indian Ocean would pass over the subcontinent, leaving India a desert. Many great rivers that nourish the land, including the Indus, the Ganga, and the Brahmaputra, originate in the Himalayas. More than 50 million people now live in the region.

The Himalayas are spread over 6,00,000 sq. km in a broad arc 2,500 km long between the Indus and the Brahmaputra. The forest cover was once very thick, with flora and fauna varying extensively from one region to the other. One-tenth of the world's known species of higher altitude plants and animals occur in the Himalayas.

Ecological Destruction

Today, however, the paradise is in danger. Human activities like forest clearing, road construction, and mountaineering have taken a heavy toll. These, in turn, have led to soil erosion, landslides, and floods. In addition, there are natural calamities like avalanches and earthquakes.

With the growth of towns in the region, the commercial felling of trees has increased to unsustainable levels. Deforestation continues in spite of the resistance by the local people, creation of protected areas, and ban orders by the government. According to scientists, the region had lost 15 per cent of its forest cover between 1970 and 2000. By 2100, it will have lost 50 per cent of its forests.

Mountaineering and trekking teams leave behind enormous amounts of garbage. They also consume large amounts of local vegetation as fuel wood and fodder. Their movements also disturb the wildlife in the mountains.

A dense network of roads has been built in the Himalayas. While they improve access for the local people, they also make it easy for outsiders to come in and exploit the forest resources. The very act of road construction using methods like blasting disturbs the ecosystem and leads to landslides and loss of vegetation. The debris also damage agricultural fields and human settlements.

The quarrying of stones for the construction of roads and the building of houses for the rising population have been increasing in scale. This has caused loss of vegetation and topsoil, lowered the water table, disturbed wildlife, and increased air pollution. Moreover, the government is setting up many hydropower projects in the region without adequate studies on the environmental impact.

Melting Glaciers

The Himalayan ecosystem is now threatened by the biggest danger of all: the huge glaciers are melting. Global warming, caused by human activities such as the burning of fossil fuels, is the most likely cause.

There is enough scientific evidence to show that glaciers are melting all over the world. Glaciologists have found that many of the Himalayan glaciers, particularly those on the Tibetan side, are retreating. However, we do not know when the Himalayas will become ice-free.

The eastern Himalayas have the largest concentrations of glaciers outside the polar region, with nearly 33,000 sq. km of glacier coverage. They feed seven of Asia's greatest rivers—Ganga, Indus, Brahmaputra, Salween, Mekong, Yangtze, and Huang He—which ensure a year-round water supply to millions of people.

Box 1.1 Find Out and Be Surprised!

The Himalayan glaciers supply water to a large number of people in the region. Can you guess what percentage of the world population does this cover?

Answer in *Appendix C*

If and when the Himalayan glaciers melt, the consequences will be serious. Initially, it will cause a rise in river levels leading to bigger floods and landslides. In the long term, however, as the volume of ice diminishes, a reduction in river flows will ensue. This, in turn, will have disastrous repercussions for India, Pakistan, Nepal, Bangladesh, and China.

In spite of their scale and grandeur, the Himalayas constitute a fragile ecosystem in delicate balance. That balance may have already been upset, with unpredictable consequences for the entire subcontinent.

Do you find it difficult to believe that the great Himalayan range could be so threatened by human activities? Do not underestimate how much harm our activities can do to nature. Very few areas of the world have been left untouched by man.

That was the story of the decline of a large ecosystem. Let us now look at the state of a smaller region in rural India.

> **Box 1.2 *Hopeful Signs: The Himalayas***
>
> Environmentalists, activists, international agencies, and the governments of the region are now working to save the Himalayan environment. As part of the National Action Plan on Climate Change (NAPCC), there is a National Mission for Sustaining the Himalayan Ecosystem. The aim is to conserve biodiversity, forest cover, and other ecological values in the Himalayan region.

THE STATE OF A DISTRICT: PROSPERITY THEN, POVERTY NOW

The monsoon has rarely failed this area. It gets an average annual rainfall of 1,250 mm, more than what Punjab receives. The water table in some places is very high. Yet, Kalahandi in western Orissa is frequently in the news for its extreme poverty and deprivation. Often it is under drought and sometimes it has floods. The people migrate to all parts of India looking for work and survival.

Kalahandi, however, was not always a place of hunger and deprivation. Nineteenth-century travellers have talked about a mass of jungles and hills in the region. Just a few decades ago, it was all green

BANGALORE

KEEP OUR EARTH GREEN
SAVE TREES
SAVE TREES
SAVE OUR EARTH
PLANT A TREE
KALAHANDI
AM BOTTLING HIMALAYAN WATER. WILL SELL THEM AT A HIGH PRICE WHEN THE RIVERS RUN DRY.
MINERAL WATER
HOW DID YOU GET HERE EARLIER THAN ME?
YOU DROVE YOUR CAR. I WALKED.

here. The forests provided livelihood for six months in a year and agriculture for the other six.

In the past, there was a network of about 30,000 traditional water-harvesting structures in the area. They included ponds, lakes, check-dams, and even tanks within paddy fields. The whole system was designed to suit the topography of the land so that no part of the rainwater went waste. What is more, the system was under the control of the community, which ensured proper maintenance and water-sharing through *jal-sabhas* (water councils).

Many tanks were built with the free labour given by the people. There was also a social system to protect the forests in the catchment areas. About 50 per cent of the cultivated land was irrigated using the water bodies. Failure of rainfall occurred but there was never any scarcity of water. With a large diversity of crops, it was one of the richest areas in eastern India. How did things change?

Decline of the Ecosystem

The troubles had started even during the British rule, but the conditions worsened after Independence. First, the government took over many of the structures, but did not maintain them properly. In other cases, fearing a takeover, the landowners converted the tanks into croplands. The focus shifted to large irrigation projects like the Hirakud Dam, though the canal system often did not suit the local topography.

Concurrently, forests were being cut down for timber and the resulting erosion of the soil dumped a lot of silt on the catchment areas. Riverbeds also silted up, leading to floods downstream. A large amount of water was being lost as run-off. From then on, even a slight shortfall in rain brought water scarcity and large-scale crop failure. Agriculture became a difficult proposition and people started migrating in large numbers, affecting community life and preventing any revival of the water harvesting system. Kalahandi thus slid into a vicious cycle, from which it has never recovered.

You have seen so far the fate of a large ecosystem and that of a smaller rural region. Let us now turn to a large city.

THE STATE OF A CITY: GARDEN CITY THEN, GARBAGE DUMP NOW

Until recently, Bangalore in Karnataka was a city of over 400 lakes, large parks, and tree-lined avenues. The lakes supplied all the drinking water needed by the population. The parks provided lung space and the trees ensured a pleasant climate. Vehicles were not many, there were no traffic jams, and the air was clean.

Over the past 15–20 years, however, this 'garden' city has seen a rapid decline of its environment. Many lakes disappeared due to encroachment and development. Others were filled with the city's garbage. Just about 60 lakes are now left and even these are heavily polluted. Thousands of trees have been cut for widening roads and constructing buildings.

The municipal water supply is pumped from the river Cauvery over a distance of 95 km. The available river water is limited and it cannot meet the ever-increasing demand of the city's population. The current supply meets just about half the demand and the rest comes from thousands of borewells. The water table is dropping rapidly and much of the groundwater is also contaminated.

There are more than 32 lakh vehicles on the roads and 1,500 new vehicles are added every day. The traffic on most of the main roads is two to three times the original capacity. Building more and more flyovers and underpasses and widening the roads only provide temporary relief, since the number of vehicles is increasing all the time.

Vehicle exhaust and industrial gases have led to severe air pollution in the city. The particulate matter in the air is much higher than the permissible level. As a result, more and more people are suffering from respiratory problems.

The city generates about 3,500 tonnes of municipal garbage every day. The two main dumping yards can only take 800 tonnes a day. The contractors are thus forced to dump the garbage wherever they find some space.

About 10 per cent of solid waste generated in the city is non-degradable plastic—mostly carry bags, sachets, and bottles. A good part of this plastic is never collected. Such items choke drains and worsen the monsoon floods. Animals eat them and fall ill.

The rapid economic growth of the city increased the demand for office space and apartments. Builders acquired vacant spaces, demolished old houses, and built tall towers, sometimes even on former lakebeds. All these residential and office blocks need water, power, and sewage facilities.

With lakes and wells gone and most of the land paved with concrete or asphalt, rainwater has nowhere to go. One sharp shower results in the flooding of roads and residential areas, especially those built on former lakebeds.

Thousands of trees were removed to make space for the city's development. As green spaces shrink and water bodies and wetlands disappear, many species of birds, animals, and reptiles are becoming locally extinct. For example, house sparrows, spotted doves, waterfowls, and vultures are rapidly declining in numbers.

Power consumption in the city has seen an enormous increase. Bangalore now uses about 35 per cent of the total power generated in the state. Air-conditioned office blocks, towering apartments, and big malls are power guzzlers. The available supply cannot meet the demand and power cuts are frequent and unpredictable.

On the whole, Bangalore can no longer be called 'a garden city'.

WHAT DO THE THREE STORIES TELL US?

The cases covered a large ecosystem, a rural area, and a large city. All have suffered environmental decline. Note three aspects of this decline:

- The destruction of the environment has taken place very rapidly, over the last few decades.
- Most of the environmental problems have been caused by human activities.
- The decline has occurred even as India registered rapid economic growth and became a major power in world economy.

Box 1.3 *Hopeful Signs: Bangalore*

In Bangalore, city officials and concerned citizens' groups are trying to improve matters. In recent years, the bus transport system has improved. Many commuters now travel by the new air-conditioned buses, giving up their cars. Others are joining car pools.

NGOs have filed cases against indiscriminate tree felling and saved some of the trees. The government is planting hundreds of trees. Many parks have been renovated and new ones constructed. The municipal corporation has made rainwater harvesting and solar water heaters compulsory for all houses.

India is, of course, not unique in experiencing environmental destruction. There are similar stories in all parts of the world. Glaciers are melting rapidly in the polar regions and in many mountain ranges. Huge forests such as those of the Amazon are being cut down. Every major city in the developing countries faces serious problems.

Clearly, human activities have led to the degradation of air and land, decline of biological diversity, scarcity of water, increasing waste, and so on. Looming over these problems are the issues of global warming and climate change that could have serious consequences. What is more, we have very little time in which to take action and reverse the decline.

WHAT IS THIS BOOK ABOUT?

In this book, we will discuss the major environmental issues. We will cover the bad news, but we will also look for signs of hope and scope for individual action.

Box 1.4 What You Should Know about the Global Environment

The key messages of this book are:

- The natural resources of our planet are limited.
- Unceasing economic development and industrial growth have led to the overexploitation of the natural resources.
- We are using up natural resources that rightfully belong to our children, grandchildren, and the future generations.
- As we exploit the resources, we are also rapidly and heavily polluting the planet.
- The current model of economic growth is clearly unsustainable in a finite world and we face environmental and social destruction.
- We can still save the planet, but we have to begin acting now.
- While each one of us is part of the problem, each of us can also become part of the solution and make a difference.
- You must begin by taking action in your lives and in your spheres of influence to heal the earth.

END PIECE

There is the story of the man who read that smoking was bad. So he gave up reading! Don't quit; this book will show you the crisis as well as the cure.

EXPLORE FURTHER

Books and Articles

Mahapatra, Richard and Ranjan Panda, 2001, 'The Myth of Kalahandi: A Resource-rich Region Reels under a Government-induced Drought', *Down to Earth*, Vol. 9, No. 21 (31 March).

Sainath, P., 1996, *Everybody Loves a Good Drought: Stories from India's Poorest Districts*, New Delhi: Penguin Books India. (Includes two stories on Kalahandi.)

Websites

International Centre for Integrated Mountain Development (ICIMD), Kathmandu, Nepal: www.icimod.org

Janaagraha, an institution in Bangalore for citizenship and democracy: www.janaagraha.org

Mountain Forum, a global network: www.mtnforum.org

THE CRISIS AND ITS CAUSES

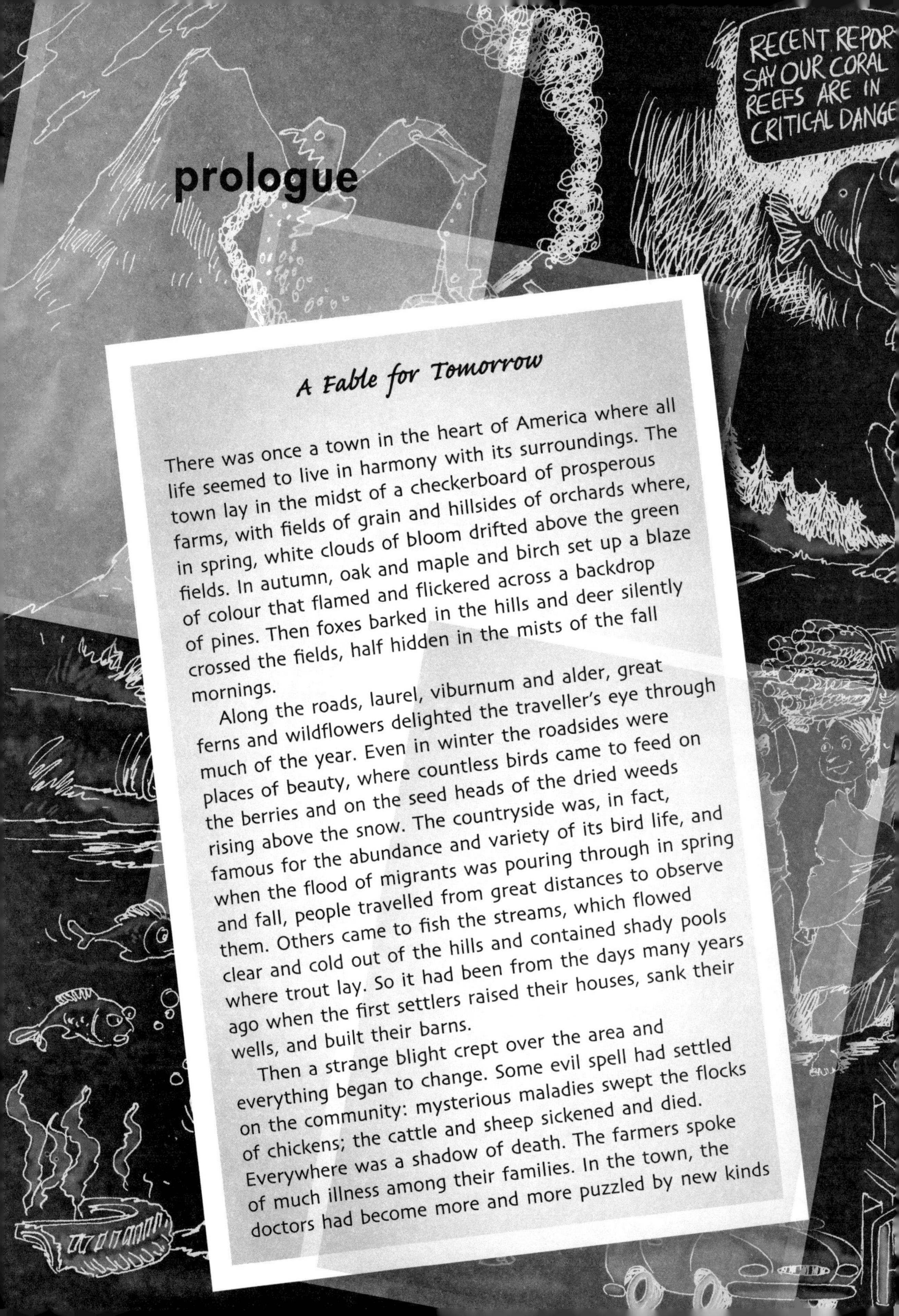

prologue

A Fable for Tomorrow

There was once a town in the heart of America where all life seemed to live in harmony with its surroundings. The town lay in the midst of a checkerboard of prosperous farms, with fields of grain and hillsides of orchards where, in spring, white clouds of bloom drifted above the green fields. In autumn, oak and maple and birch set up a blaze of colour that flamed and flickered across a backdrop of pines. Then foxes barked in the hills and deer silently crossed the fields, half hidden in the mists of the fall mornings.

Along the roads, laurel, viburnum and alder, great ferns and wildflowers delighted the traveller's eye through much of the year. Even in winter the roadsides were places of beauty, where countless birds came to feed on the berries and on the seed heads of the dried weeds rising above the snow. The countryside was, in fact, famous for the abundance and variety of its bird life, and when the flood of migrants was pouring through in spring and fall, people travelled from great distances to observe them. Others came to fish the streams, which flowed clear and cold out of the hills and contained shady pools where trout lay. So it had been from the days many years ago when the first settlers raised their houses, sank their wells, and built their barns.

Then a strange blight crept over the area and everything began to change. Some evil spell had settled on the community: mysterious maladies swept the flocks of chickens; the cattle and sheep sickened and died. Everywhere was a shadow of death. The farmers spoke of much illness among their families. In the town, the doctors had become more and more puzzled by new kinds

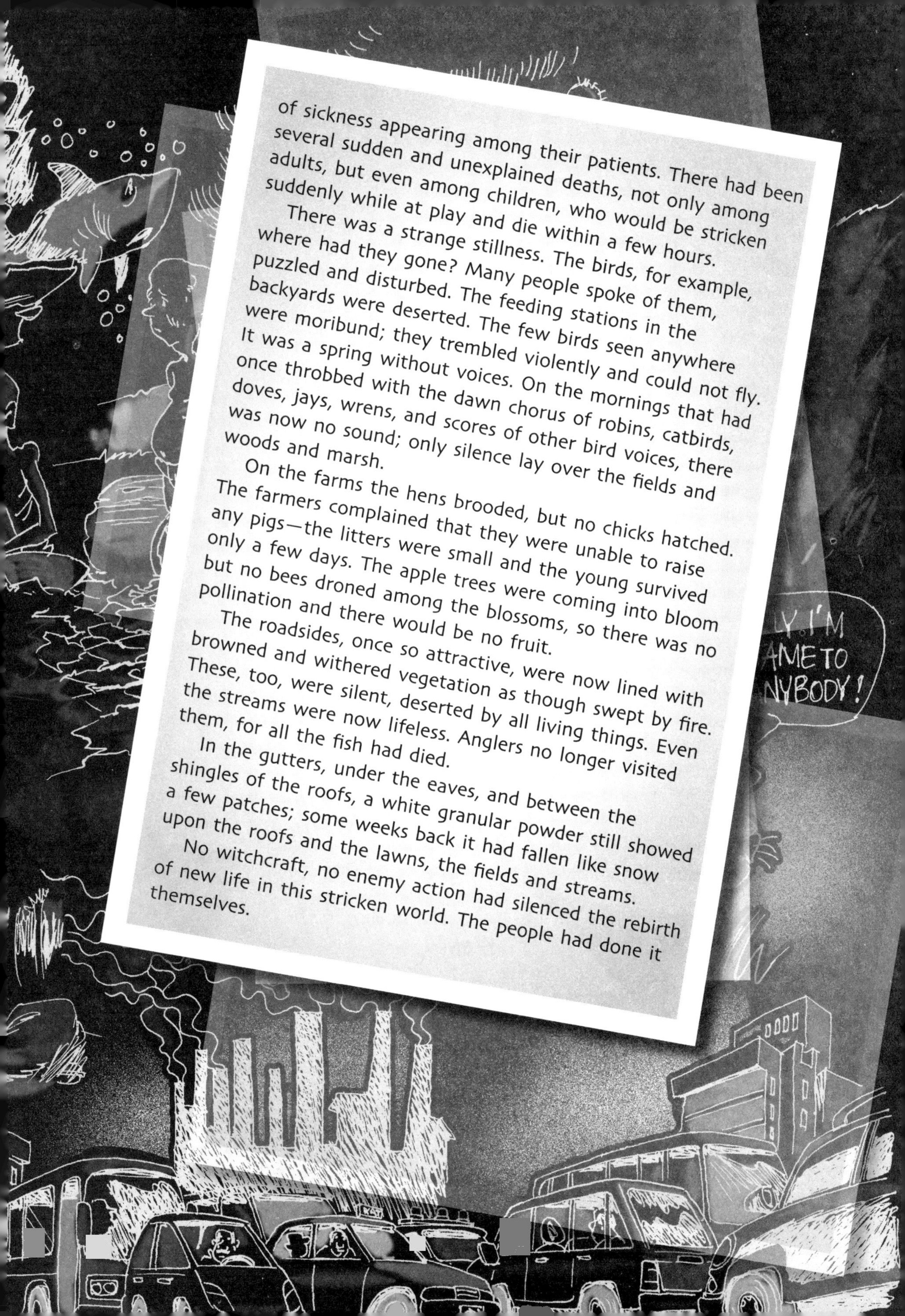

of sickness appearing among their patients. There had been several sudden and unexplained deaths, not only among adults, but even among children, who would be stricken suddenly while at play and die within a few hours.

There was a strange stillness. The birds, for example, where had they gone? Many people spoke of them, puzzled and disturbed. The feeding stations in the backyards were deserted. The few birds seen anywhere were moribund; they trembled violently and could not fly. It was a spring without voices. On the mornings that had once throbbed with the dawn chorus of robins, catbirds, doves, jays, wrens, and scores of other bird voices, there was now no sound; only silence lay over the fields and woods and marsh.

On the farms the hens brooded, but no chicks hatched. The farmers complained that they were unable to raise any pigs—the litters were small and the young survived only a few days. The apple trees were coming into bloom but no bees droned among the blossoms, so there was no pollination and there would be no fruit.

The roadsides, once so attractive, were now lined with browned and withered vegetation as though swept by fire. These, too, were silent, deserted by all living things. Even the streams were now lifeless. Anglers no longer visited them, for all the fish had died.

In the gutters, under the eaves, and between the shingles of the roofs, a white granular powder still showed a few patches; some weeks back it had fallen like snow upon the roofs and the lawns, the fields and streams.

No witchcraft, no enemy action had silenced the rebirth of new life in this stricken world. The people had done it themselves.

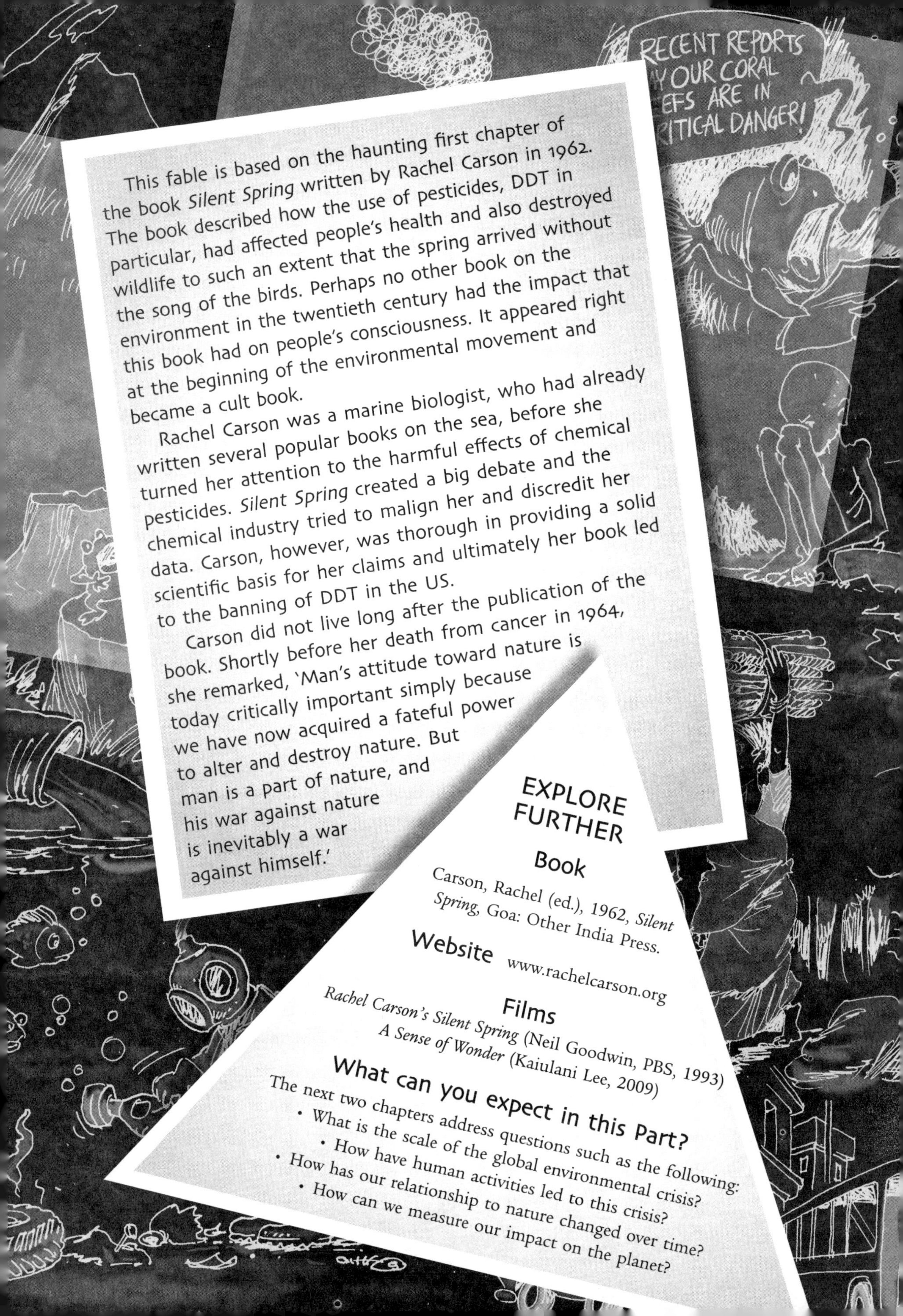

This fable is based on the haunting first chapter of the book *Silent Spring* written by Rachel Carson in 1962. The book described how the use of pesticides, DDT in particular, had affected people's health and also destroyed wildlife to such an extent that the spring arrived without the song of the birds. Perhaps no other book on the environment in the twentieth century had the impact that this book had on people's consciousness. It appeared right at the beginning of the environmental movement and became a cult book.

Rachel Carson was a marine biologist, who had already written several popular books on the sea, before she turned her attention to the harmful effects of chemical pesticides. *Silent Spring* created a big debate and the chemical industry tried to malign her and discredit her data. Carson, however, was thorough in providing a solid scientific basis for her claims and ultimately her book led to the banning of DDT in the US.

Carson did not live long after the publication of the book. Shortly before her death from cancer in 1964, she remarked, 'Man's attitude toward nature is today critically important simply because we have now acquired a fateful power to alter and destroy nature. But man is a part of nature, and his war against nature is inevitably a war against himself.'

EXPLORE FURTHER

Book

Carson, Rachel (ed.), 1962, *Silent Spring*, Goa: Other India Press.

Website

www.rachelcarson.org

Films

Rachel Carson's Silent Spring (Neil Goodwin, PBS, 1993)

A Sense of Wonder (Kaiulani Lee, 2009)

What can you expect in this Part?

The next two chapters address questions such as the following:

- What is the scale of the global environmental crisis?
- How have human activities led to this crisis?
- How has our relationship to nature changed over time?
- How can we measure our impact on the planet?

2 THE GLOBAL ENVIRONMENTAL CRISIS

Don't worry about the world coming to an end today.
It is already tomorrow in Australia.

—CHARLES M. SCHULTZ*

THE STORY OF EASTER ISLAND: CUT THE TREES, DESTROY A SOCIETY

On the Easter of 1722, a group of Dutch explorers arrived at a small, isolated island in the South Pacific and named it the Easter Island. They found 2,000 inhabitants struggling to live on the barren island. The explorers heard from the people their strange story. Later, archaeologists and historians pieced together what happened there.

Easter Island, about 160 sq. km in area, was first colonized 2,500 years ago by the Polynesians. The settlers were almost completely dependent on the island's palm trees. The trees gave the people many things: fuel, food, houses, tools, boats, rope, and clothing.

By AD 1400 the population had increased to almost 20,000. The people were happy and had even developed arts and crafts. They sculpted a large number of big stone statues and moved them to the shore. Many of these are still standing.

The islanders, however, made a fatal blunder. They did not conserve the palm trees. They cut and used up the trees so fast that soon the last tree disappeared.

*(1922–2000); American cartoonist and creator of the comic strip, *Peanuts*.

They could no longer build boats to catch fish. Without any forests to absorb and retain water, springs and streams dried up, soil began to erode, food crops dwindled, and famine set in. The starving people started fighting among themselves. Their society collapsed. Easter Island is a grim reminder of what happens if we do not care for trees and other natural resources.

You might say, 'The Easter Island story happened long ago. It is a small island any way. I cannot believe that such things happen now. Given all the knowledge that we have, no society would destroy all its natural resources.'

That is a normal response to the Easter Island story. Human beings, however, seldom take lessons from history. Here is a more recent story.

THE STORY OF MADAGASCAR: HUMANS ARRIVE, FORESTS DISAPPEAR

Madagascar, the fourth largest island in the world, lies off the eastern coast of Africa. It had an astonishing biodiversity, because it existed in near isolation for 40 million years. What is more, 85 per cent of the species are endemic to the island.

However, since humans came to the island about 1,500 years ago, 80 per cent of the tropical seasonal forests and 65 per cent of the rainforests have been destroyed for lumber, fuel wood, and conversion to cropland. With forests gone, soil erosion has become severe. The rapid population growth has led to slash-and-burn agriculture on these poor soils.

Most of the biological species are now endangered. Of the 31 primates inhabiting Madagascar, 16 face extinction. There is also rampant illegal smuggling and export of endangered frogs, chameleons, and lizards. It is expected that half of all the plant and animal species in Madagascar will disappear by 2025.

Some international efforts are being made to save Madagascar's biodiversity, but chances of success are slim. Madagascar is perhaps another Easter Island in the making. To make matters worse, the island has also been facing political unrest in recent times.

Your response could well be, 'I agree that Easter Island and Madagascar show the stupidity of human beings. Yet, these places are far away. What about India? Are such things happening here?'

Let us turn to India now.

THE STORY OF THE GANGA: HOLY RIVER, UNHOLY POLLUTION

One day in the 1980s, someone dropped a burning match into the Ganga near Haridwar by chance and a whole kilometre stretch of river caught fire. The effluents from two factories released into the river were so toxic that the water caught fire. In fact, the fire went seven metres high and could not be extinguished for three hours.

The Ganga has its source in the Himalayas, flows 2,500 km across the plains of northern India, and empties into the Bay of Bengal. The Ganga and its tributaries cover a length of 12,700 km and they pass through nearly 100 cities and towns. There are millions of people dependent on the river.

The major rivers of India such as the Brahmaputra, Godavari, and Krishna are heavily polluted, but the Ganga is the worst case. According to the Central Pollution Control Board (CPCB), over 40 per cent of the Ganga's length is moderately or severely polluted.

Every day, an estimated 2 million people take a dip in the Ganga. Yet dead bodies of human beings and animals are a common sight in the river. Hundreds of factories dump their toxic waste into the river. The leather factories of Kanpur, for example, let out waste containing chromium and other chemicals.

About one billion litres of untreated sewage is dumped every day into the Ganga from the towns on its banks. Such pollution causes diseases like dysentery, typhoid, cholera, and cancer.

In 1985, environmental lawyer M.C. Mehta filed a case in the Supreme Court of India against all the industries and the municipal towns that were polluting the river. The court ordered the factories to clean their waste before letting it out into the river. Meanwhile, the Government of India initiated the Ganga Action Plan (GAP) to clean the river at a cost of Rs 10 billion. In spite of these actions, the Ganga remains very polluted. At many places, its water is unfit for drinking, washing, bathing, or irrigation.

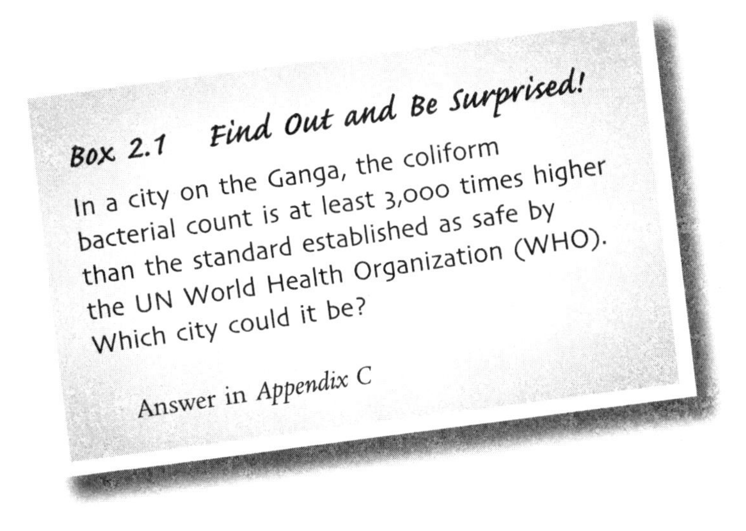

RECENT REPO
SAY OUR CORAL
REEFS ARE IN
CRITICAL DANG
GREEN PEACE

ACTUALLY I'M TOO TAME TO HURT ANYBODY!

Box 2.2 *Hopeful Signs: The Ganga*

- The Sankat Mochan Foundation (SMF) of Varanasi is fighting a great battle to save the Ganga. Set up by Veer Bhadra Mishra, a priest and a civil engineer, the foundation has studied the pollution problem and has come up with possible solutions. Mishra has created greater awareness of the state of the Ganga among the public and the bureaucracy. He is also trying to get the wastewater treated before it is released into the river. Many other individuals and groups are also trying to save the Ganga.
- In October 2009, the Government of India (GoI) formed the National Ganga River Basin Authority (NGRBA). The authority will spend Rs 150 billion over 10 years to set up infrastructure for treating sewage and wastewater. The project will concentrate on treatment of catchment area as well on river front development for effective results. No untreated industrial effluent and municipal sewage will be allowed to flow in the river after 2020.

IS THERE AN ENVIRONMENTAL CRISIS?

After reading the stories of environmental destruction in this chapter and the previous one, you might respond, 'These are local examples of decline. But the world is large and things may be balancing out over the earth. For example, trees may be lost in one place, but many more trees may be growing elsewhere. Some species may be going extinct, but we also discover new species. Is there really a global environmental crisis?'

There are differing perceptions about the problems of the environment. Here are some views:

- Environmental problems are not as bad as they are made out to be.
- There are problems, but science and technology will find solutions.
- There is a real crisis, the wolf is at the door, there is no way out!
- For the developing countries, hunger and poverty are more important issues than the environment. They need quick economic growth.
- It is all part of a 'grand design'. God made this world, and He will take care of it.

- The earth is a self-evolving and self-regulating living system and it will survive, even if humans destroy themselves.

By the time you finish reading this book, you will perhaps form your view.
How can you find out about the state of the global environment? You can turn to many reliable and regular reports such as:

- *Global Environment Outlook*, compiled every three years or so by the United Nations Environment Program (UNEP).
- *Living Planet Report*, released every two years by the World Wide Fund for Nature (WWF).
- *State of the World Report*, published annually by the Worldwatch Institute, Washington, DC, USA.
- *State of India's Environment*, a series of reports from the Centre for Science and Environment (CSE), New Delhi.

There are also other sources of information such as the International Union for Conservation of Nature (IUCN), World Resources Institute, Greenpeace, and Earth Policy Institute. In addition, you can read periodical and thematic assessments such as the UN World Water Development Report and periodical assessments by the Intergovernmental Panel for Climate Change (IPCC). Other examples are the Report of the Independent World Commission on Dams (2000) and the UN Global Millennium Ecosystem Assessment (2005). (See Boxes 2.3 and 2.4.)

WHAT DO THE REPORTS SAY?

Here are some indicators of the current state of the planet, drawn from the various reports:

Population

- The world population is expected to reach 7 billion in 2011. It has taken just 12 years for the population to increase from 6 to 7 billion.
- India's population was about 1.2 billion in 2010. By 2030, we will overtake China in population.

Water and Sanitation

- Two billion people live in countries that are water-stressed and, by 2050, four billion people could be suffering from water stress. Half the world population today lacks sanitation facilities.
- In India, more than 60,000 villages are without a single source of drinking water. Diarrhoea, brought on by contaminated water, claims the lives of one million children every year. In addition, 45 million people are affected annually by the poor quality of water.

Biological Diversity

- Worldwide, 24 per cent of mammals, 12 per cent of birds, 25 per cent of reptiles, and 30 per cemt of fish species are threatened or endangered; this is 100 to 1,000 times the rate at which species naturally disappear.
- More than 10 per cent of India's recorded wild flora and fauna are threatened and many are on the verge of extinction.

Box 2.3 *Ecosystems and Ecosystem Services*

When you hear the term 'forest ecosystem', what comes to your mind? Surely you think immediately of trees, animals, birds, butterflies, and so on. The mention of the term 'ocean ecosystem' brings to mind a picture of waves, currents, fish, whales, and so on. Similarly, we talk about a 'desert ecosystem' or 'a mountain ecosystem'. An ecosystem (or ecological system) is a region in which living organisms interact with their environment.

Ecosystem services are the benefits people obtain from ecosystems. Nature provides these services to us free of cost. They are of four types:

- Provisioning services: The products obtained from ecosystems, including genetic resources (like seeds), wood, food, and freshwater.
- Regulating services: The benefits obtained from the regulation of ecosystem processes, including the regulation of climate, water purification, flood control, and control of human diseases.
- Cultural services: The non-material benefits people obtain from ecosystems including education, recreation, and experiencing the beauty of nature.
- Supporting services: Ecosystem services that are necessary for the production of all other ecosystem services. Some examples are biomass production, production of atmospheric oxygen, and soil formation.

Can we put a monetary value on ecosystem services? Scientists estimate that nature's services are worth more than USD 36 trillion per year. This is comparable to the annual Gross World Product, that is, the value of all the goods and services we produce. Other experts believe that this is a conservative figure and that the real worth of nature could be much more, even a million times more.

Forests

- The net loss in global forest area during the 1990s was about 94 million ha. Tropical forests are being cleared at the rate of 70,000 to 170,000 sq. km annually (equal to 21–50 soccer fields per minute)
- India's forest cover declined from 40 per cent a century ago to 22 per cent in 1951 and to 19 per cent in 1997. The quantitative decline is supposed to have been arrested since 1991, but the qualitative decline persists.

Box 2.4 UN Millennium Ecosystem Assessment

The UN Millennium Assessment, made between 2000 and 2005 by a large team of world scientists, addressed the following questions:

- How have the world's ecosystems changed over the past 50 years?
- What has been the impact on ecosystem services, that is, the benefits people obtain from ecosystems?
- How have changes in ecosystem services affected human wellbeing?
- How could ecosystem changes affect people in future decades?
- What should we do at local, national, or global levels to improve ecosystem management and thereby improve human well-being and alleviate poverty?

The main findings of the assessment were:

- Over the past 50 years, we have changed ecosystems more rapidly and extensively than in any comparable period of time in human history. We have done so mainly to meet our rapidly growing needs for food, freshwater, timber, fibre, and fuel.
- The changes that we have made to ecosystems have improved human well-being and speeded up economic development. But they have also led to the degradation of most ecosystems and, hence, affected many ecosystem services.
- The degradation of ecosystems could grow significantly worse during the first half of this century.
- We can reverse the degradation of ecosystems to some extent, but we have to make many changes in our activities. Such changes are not happening now.

Land

- Each year, six million ha of agricultural land is lost due to desertification and soil degradation. This process affects about 250 million people in the world.
- India has nearly 130 million ha of wasteland compared to 305 million ha of biomass producing area.

Pollution

- At least one billion people in the world breathe unhealthy air and three million die annually due to air pollution.
- The World Health Organization (WHO) consistently rates New Delhi and Kolkata as being among the most polluted megacities of the world.

Coastal Areas and the Ocean

- Worldwide, 50 per cent of mangroves and wetlands that perform vital ecological functions have been destroyed. About 80 per cent of world fish stocks are depleted, fully exploited, or overexploited.
- Over the past 40 years, India too has lost more than 50 per cent of its mangrove forests. The absence of mangroves on the Orissa coast accentuated the damage to life and habitation when the Supercyclone struck in 1999.

Disasters

- Across the world, the numbers of people affected by disasters have risen from an average of 150 million a year in the 1980s to 200 million a year in the 1990s. Floods, droughts, and windstorms accounted for more than 90 per cent of the people killed in natural disasters.
- India is the most disaster-prone country in the South Asian region. Drought, floods, earthquakes, and cyclones occur with grim regularity. In the Orissa supercyclone of 1999, 10,000 people died; 16,000 died in the Gujarat earthquake in 2001. In 2010, there were heavy floods in many parts of north India. Again in 2010, severe floods occurred in Pakistan, affecting more than 20 million people.

Energy

- More than two billion people in the world go without adequate energy supplies. Fossil fuels are being overexploited and oil prices are rising.
- India imports about 70 per cent of its oil needs, primarily to feed the transportation sector.

Box 2.5 Why We Ignore the Environmental Crisis

If so many reports by the UN and other agencies paint a bleak picture of the environment, why are we ignoring them? Why are people and governments not taking action to reverse the decline of the environment? Here are some reasons:

- Lack of awareness: Our information about the world comes mainly from the media such as newspapers and TV, which focus on politics and sensational stories. They give very little importance to environment-related news, though this is now changing.
- Disinformation: Vested interests that want to maintain the status quo often spread wrong information to confuse the public. For example, companies launch misleading advertisements to forestall environmental regulations that would increase their costs or affect their business.
- Lack of political will: Governments have a fixed tenure and, with elections in mind, they do not want to take unpopular measures.
- Feeling of powerlessness: Faced with doomsday scenarios, the average citizen feels helpless and tends to deny the very existence of problems.
- Beliefs: Many have faith that either science or God will solve the problems and save the earth.

Global Warming and Climate Change

- There are clear signs of global warming and climate change: global temperatures in 2010 were highest on record; permafrost and glaciers in the polar and other regions are melting; the very existence of many small islands is being threatened by sea level rise.
- In India, the rainfall patterns and climate are becoming unpredictable. The year 2010 was the warmest in India since 1901. Even more striking is the fact that, of the ten warmest years since 1901, seven were recorded since 2002. There is also evidence to show that some of the Himalayan glaciers are losing mass and becoming thinner.

Urbanization

- More than half the world's population now lives in urban areas, compared to little more than one-third in 1972. About one-quarter of the urban population lives below the poverty line.
- About 23 per cent of the population in India's million-plus cities lives in slums. Dharavi in Mumbai, the largest slum in Asia, houses 5,00,000 people over a small area of 170 ha.
- Major Indian cities do not meet our National Ambient Air Quality Standards (NAAQS).

This is just a sample of the environmental indicators. We will discuss each aspect of the environment in Part II of the book.

The basic facts of the environmental crisis are certainly depressing, but we cannot run away from them. When we take up each of the environmental topics, we will suggest positive actions that you can take to save the planet.

Box 2.6 For the Future

Planting trees early in spring,
we make a place for birds to sing
in time to come. How do we know?
They are singing here now.
There is no other guarantee
that singing will ever be.

—Wendell Berry*

*b. 1934; American farmer, writer, and environmentalist.

END PIECE

Let us close with a Chinese saying:

If you would be happy for a while, take a wife; if you would be happy for a month, kill a pig; but if you would be happy all your life, plant a garden.

EXPLORE FURTHER

Books and Articles

Agarwal, Anil, Sunita Narain, and Srabani Sen (eds), 1999, *State of India's Environment—5, Fifth Citizens' Report, Part I: National Overview* and *Part II: Statistical Database*, New Delhi: Centre for Science and Environment.

Gonick, Larry and Alice Outwater, 1996, *The Cartoon Guide to the Environment*, New York: HarperCollins Publishers.

Hammer, Joshua, 2007, 'A Prayer for the Ganges: Across India', *Smithsonian Magazine* (November).
Orliac, Catherine and Michel Orliac, 1995, *The Silent Gods: Mysteries of Easter Island*, London: Thames and Hudson.
Somerville, Richard C. J., 2008, *The Forgiving Air: Understanding Environmental Change*, second ed., Washington, D.C: American Meteorological Society.
UNEP, 2001, *India: State of the Environment 2001*, Bangkok: United Nations Environment Programme, Regional Resource Centre for Asia and the Pacific.
____, 2003, *Global Environment Outlook 3*, Geneva: UNEP Global Resource Information Division, Division of Early Warning and Assessment.

Websites

Eastern Island: www.mysteriousplaces.com/Easter_Island/
Ganga Pollution: http://sankatmochan.tripod.com/GangaPollution.htm
www.experiencefestival.com/a/Ganges_River_-_Recent_Pollution/id/1416835
Madagascar: www.partners4madagascar.org/ecology-of-madagascar.html
www.pbs.org/edens/madagascar/eden.htm
www.wildmadagascar.org
Scientific reports on environmental topics: www.greenfacts.org
UN Millennium Ecosystem Assessment: www.millenniumassessment.org
World Wide Fund for Nature: www.panda.org
Worldwatch Institute: www.worldwatch.org

3 WHY ARE WE IN A CRISIS?

Anyone who believes that
exponential growth can go on forever
in a finite world is
either a madman or an economist.

—Kenneth E. Boulding

ROOTS OF THE CRISIS

The previous chapters presented depressing data about the global environment. In this chapter, we will try to get to the roots of the crisis.

Human activities are certainly responsible for most of the environmental problems. What we do to nature amounts to two things:

- First, we are consuming natural resources at a rapid rate. In fact, we are doing so faster and faster. We do not give enough time for nature to regenerate its resources.
- Second, as we use up the resources, we are rapidly and heavily polluting the environment. Nature cannot absorb this pollution.

Let us examine these two points. All living things consume natural resources for their survival. As long as the rate of consumption is reasonable, nature has the capacity to regenerate the resources. For example, we consume water, but the global water cycle ensures that the earth gets freshwater every year through rain. The soil also stores water and gives it back to us. Now, however, we are using up water at such a rate that nature cannot replenish the stock of groundwater fast enough.

Nature also has ways of handling pollution. For example, when polluted water flows over a wetland, a natural cleaning action takes place. However, if the pollution is too much, or the wetland itself has been replaced by habitation, the water cannot be cleaned by the natural process.

Nature has a budget: it can only produce so much resource and absorb so much waste every year. The problem is that our demand for nature's services exceeds what it can provide.

Another way of examining the human exploitation of natural resources is through the concept of 'ecological footprint'.

ECOLOGICAL FOOTPRINT

Let us do a mental experiment. Take the geographical area of a city like Bangalore and cover it with a huge glass hemisphere. We let in sunlight, but we do not allow any material to enter or leave the enclosure. How long will the population of the city survive? Surely, the city would face serious problems just in a few days. You know how a small strike by truckers creates chaos in any city.

What will happen to Bangalore in our experiment? The city cannot produce enough food for all the people. There will also be severe water scarcity, because the supply of Cauvery water will stop. Also, the water tankers from the surrounding villages cannot enter the city.

The enormous amount of solid waste generated every day cannot be sent out to the landfills. The air trapped in the hemisphere will soon become so polluted that people will find it difficult to breathe. The 'carrying capacity' of the city area is not sufficient to sustain the lives of the population. Any city is heavily dependent on the outside world for its needs.

Let us examine that dependence further. Suppose we are able to expand the size of the glass hemisphere to take in more and more of the surrounding area. Assume also that this area has diverse natural resources like a mini-earth.

We can now ask another question: How large should be the area covered, if we want the city to survive indefinitely on the land, water, and energy resources available within the hemisphere? That area is the city's 'ecological footprint'.

The millions of people living in Bangalore have huge needs and they draw their requirements from a very large area surrounding the city. The ecological footprint of the city is many times larger than its actual area. Similarly, we can talk about the ecological footprint of a country. Or, we can compare the footprint of a citizen of US with that of an Indian citizen.

We define ecological footprint as the amount of biologically productive land and sea area required to sustain indefinitely a person, a city's population, a country, the manufacture of a product, and so on. It accounts for the use

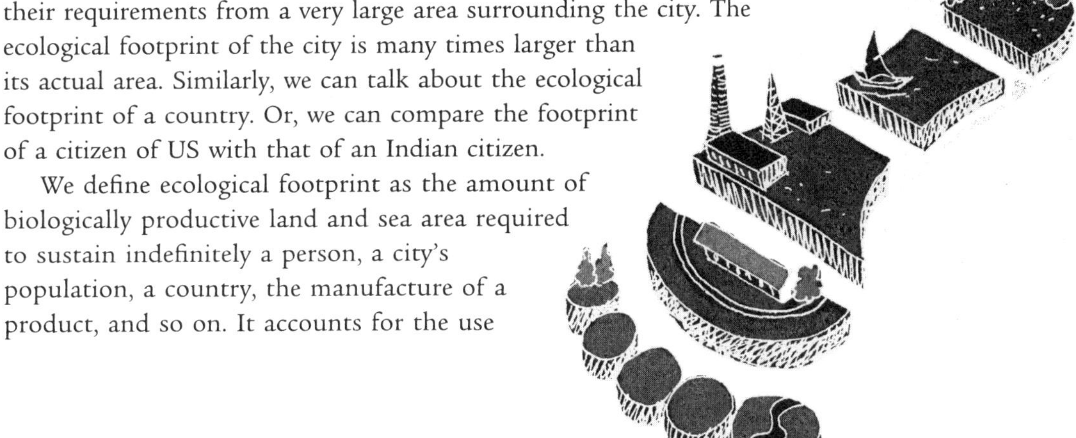

of energy, food, water, building material, and other consumables and the wastes created.

What do we mean by 'biologically productive land and sea area'? This includes the area that 1) supports human demand for food, fibre, timber, energy and space for infrastructure and 2) absorbs the waste products from the human economy. Biologically productive areas include cropland, forest and fishing grounds, and do not include deserts, glaciers, and the open ocean.

Since the productivity of land and sea differs from place to place, we use 'global hectare' as a measurement unit. This makes data and results globally comparable. A global hectare (gha) is a common unit that expresses the average productivity of all the biologically productive land and sea area in the world in a given year. We convert physical hectares of different types of land, such as cropland and pasture, into the common unit of global hectares.

We can express ecological footprint either in gha or as the ratio of the area required to the actual area of the entity. For example, if Bangalore requires for its survival an area three times its geographical spread (both areas measured in gha.), its ecological footprint is three.

How is our environmental crisis connected to the idea of ecological footprint?

The larger the footprint, the more is the consumption of natural resources and thereby environmental degradation. Most of the world's cities have footprints greater than one. Again, countries like the US have large footprints too. What then could be the ecological footprint of humanity as a whole?

HUMANITY'S ECOLOGICAL FOOTPRINT

You would probably think that humanity's footprint ought to be less than one. In fact, our footprint is already more than 1.4. That is, we require more than the earth's area to sustain humanity's consumption of natural resources!

In 2005 there were 13.4 billion ha of biologically productive land and sea available and 6.5 billion people on the planet. That is, an average of 2.1 gha was available per person. By that time, however, our global consumption of resources required an area of 2.7 gha per person. That is, humanity's ecological footprint was 1.3.

What is worse, it is increasing all the time. By 2009, the global ecological footprint was 1.4. That is, we now require the equivalent of 1.4 planets to support our lifestyles.

How is it possible? Common sense tells us that we could not be using resources from an area larger than that of the earth! We have only one earth. If our footprint is 1.4, how are we surviving at all?

USING TOMORROW'S RESOURCES

We survive, because each year we are using up more than our annual share of the earth's resources. Currently, by September or earlier, we use up ecological resources that the earth regenerates in the whole year! The rest of the year, we survive by dipping into the quota of the future. Instead of living within the 'annual interest' that nature gives us, we have begun using up our 'natural capital'. In a sense, we are using resources that rightly belong to our children and grandchildren. We are living beyond our means.

Let us return, for example, to the global water cycle. It gives us a certain amount of freshwater every year and some of it recharges our groundwater. However, we now consume (and pollute) more than this annual replenishment of water. We are thus madly exhausting our groundwater supplies. The water tables are dropping at alarming rates and soon we will not have water to pump out.

Another example is the extraction of crude oil that was formed and stored millions of years ago. We are extracting the oil so fast that the supplies will be exhausted fairly soon. Similarly, we cut trees faster than they re-grow, and catch fish at a rate faster than they repopulate.

In general, we are using resources faster than they can regenerate and creating waste faster than it can be absorbed. This is called 'ecological overshoot'. While this can be done for a short while, overshoot ultimately leads to the depletion of resources on which our economy depends.

HOW MANY EARTHS WOULD WE NEED?

Given that our footprint is already larger than the area of the earth, what will happen in the future? All the poorer countries of the world want economic development and they need natural resources. For example, India and China are developing rapidly and are consuming huge resources!

The current ecological footprints of the US, China, and India are 1.8, 2.3, and 2.2 respectively. Every citizen of the US requires 9.4 gha to meet his consumption—highest for any country. A Chinese citizen uses 2.1 gha Due to rapid economic development, China's ecological footprint has quadrupled in the last four decades.

In the same period, India's footprint has doubled. Even then, an Indian citizen needs just 0.8 gha, which is smaller than that in many other countries. At the same time, due to the large population, the country now demands the biological capacity of two Indias to provide for its consumption and absorb its wastes.

The global ecological footprint is getting larger and larger. If the world economy keeps growing at the current rate, by 2030 we will need TWO earths

BANGALORE

3R

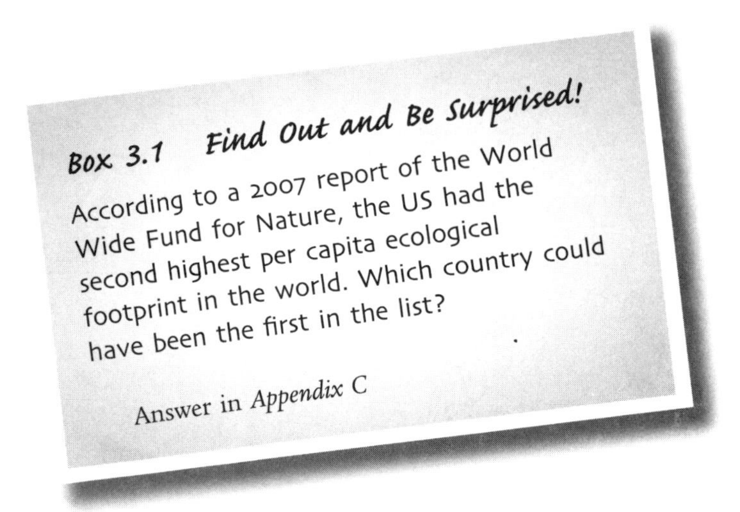

to keep up with our consumption. Alternately, if all of us were to adopt the US lifestyle, we will need FOUR earths to support us!

EXPONENTIAL GROWTH

There could be counter arguments: 'Let us look at the past. The earth is about 5 billion years old, man has been around for 3 to 5 million years, and civilization has a history of 10,000 years. We have survived for long and we have in fact mastered nature in many ways.'

'We also know that nature has the capacity to absorb disturbances and regain its balance. If there are still problems, we will solve them using the science and technology that we have developed. Why we should worry now?'

It is true we have created a civilization and made spectacular advances in science, technology, and medicine. We have established industries and created large cities and habitats. In that process, however, there has been a change in the way we carry out our activities on earth.

The Spikes

Let us look at four graphs. They show how the following four quantities have been increasing with time:

- World population
- Global consumption of goods and services
- Carbon dioxide emissions
- Number of species that go extinct each year.

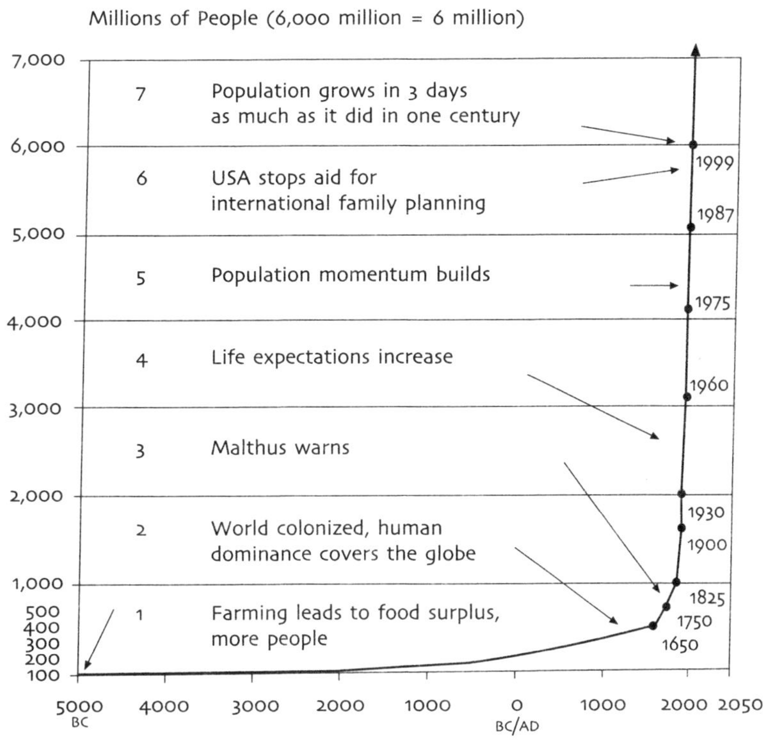

Source: Ayres (1999).

Figure 3.1 The Population Spike

First, let us look at the growth of world population over time (Figure 3.1).

You can see that the graph has the shape of a hockey stick. The mathematical name for such a graph is the exponential curve. The popular name is spike. The graph is almost flat for a long time, but at some point the quantity starts growing rapidly and the graph becomes almost vertical.

Many phenomena in the world show such exponential growth. For example, consider an investment that gets you compound interest. In the case of simple interest, the yearly interest on the capital is the same every year. If you plot the total interest earned against time, the graph will be a straight line. In the case of compound interest, the annual interest is added to the capital and the total amount earns further interest. The total interest grows like an exponential curve.

Any exponential curve has a special characteristic. The quantity doubles after a certain number of time periods. For example, the world population has been doubling every 40–45 years.

Now look at the growth of global consumption of goods and services over time (Figure 3.2).

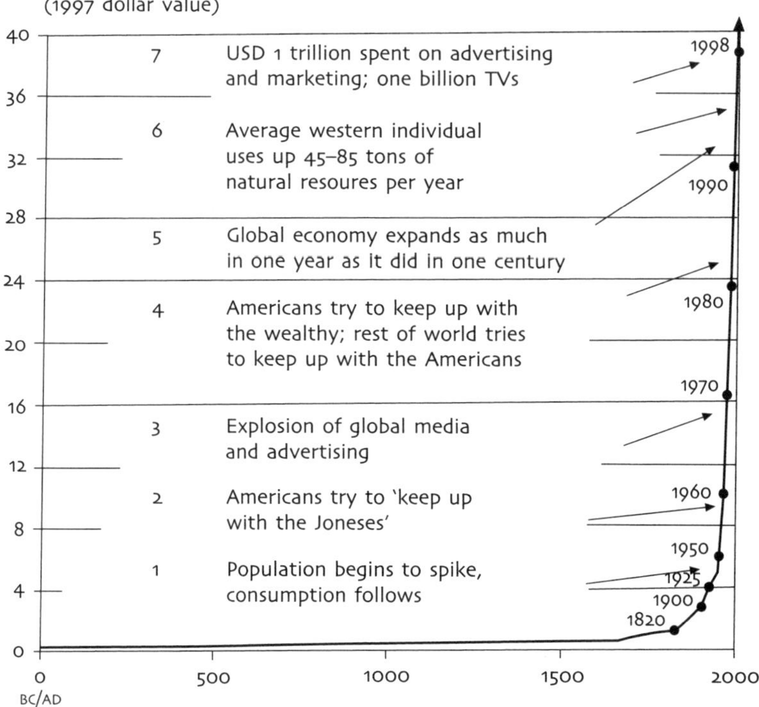

Source: Ayres (1999).

Figure 3.2 The Consumption Spike

The graph shows that the global consumption has also been growing exponentially. But that should be obvious: more the number of people, more will be the consumption.

There is, however, something that is not obvious from the graph. The consumption curve is steeper than the population one. In other words, the same people are also consuming more and more over time. It is the consumption spike that leads to excessive use of natural resources and a larger ecological footprint.

Now look at a third graph that shows the increase in carbon dioxide emissions over time (Figure 3.3).

This again is an exponential curve and the spike began around AD 1800. The spike in carbon dioxide emissions is a serious matter. As we will see in the Chapter 5, this is primarily responsible for global warming and climate change.

We now have the fourth graph, which shows the number of species that go extinct every year (Figure 3.4).

Though it is not obvious, this is the steepest of the four curves. Also, we do not normally notice the extinction of species. Human activities, however, are

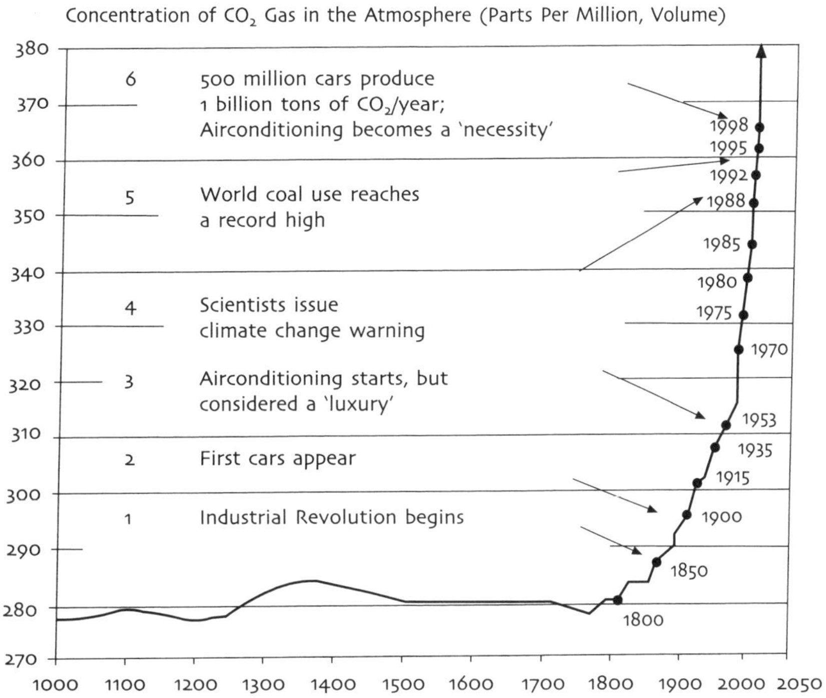

Concentration of CO_2 Gas in the Atmosphere (Parts Per Million, Volume)

6 500 million cars produce 1 billion tons of CO_2/year; Airconditioning becomes a 'necessity'

5 World coal use reaches a record high

4 Scientists issue climate change warning

3 Airconditioning starts, but considered a 'luxury'

2 First cars appear

1 Industrial Revolution begins

Source: Ayres (1999).

Figure 3.3 The Carbon Dioxide Spike

causing a mass extinction of species and this threatens the very web of life and our food security. We will take up this issue in Chapter 7.

Is there any pattern in the four graphs? It is surprising that all four show exponential growth. What is more, in each case, the spike began in recent times—200 to 400 years ago. Is there any reason for this behaviour?

The spiking began in Europe with the scientific and industrial revolutions. The advances in science and technology gave rise to a change in our attitude towards nature.

IDEA OF PROGRESS

We started believing that we were the masters of nature and that we could exploit and control nature through science and technology. Continuous economic growth through expanding trade and industry would put mankind on the path of progress. This would lead to prosperity and happiness for all.

Such an approach came to be called the 'Idea of Progress'. It is the belief that, in general, humanity moves toward improved material conditions. More and more

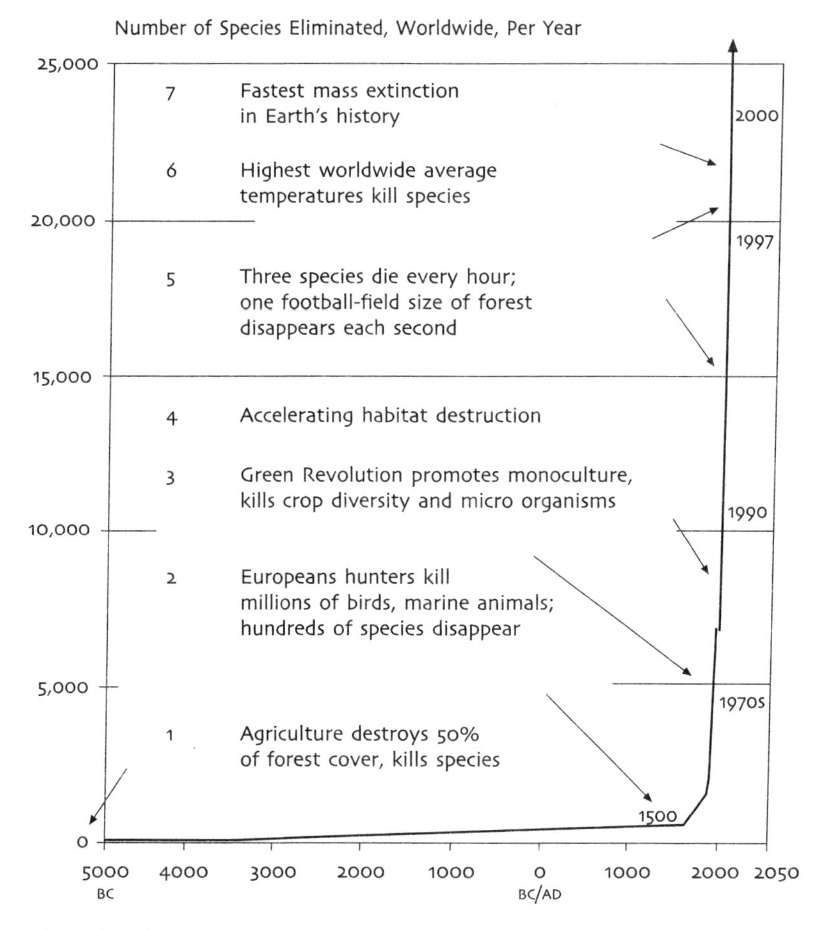

Source: Ayres (1999).

Figure 3.4 The Extinction Spike

people will have healthier, happier, more secure, and more comfortable lives. In short, the idea of progress says, 'Most things get better in the long run'.

Through colonialism and western education, this idea has spread throughout the world. Most countries, including the poorer ones, follow this path. As an educated Indian, you surely have deep and unshakable faith in progress.

Exponential growth and economic development, driven by the idea of progress, could only lead to the overexploitation of natural resources that we see today. This is clearly unsustainable in the long run. As we will see later, continuing on this path will only result in an environmental and social catastrophe.

How can we, however, find fault with economic growth, which is supposed to solve the problem of poverty? How can countries like India stop growing? Can we not balance economic development and environmental conservation?

Such concerns are valid, but it is also clear that the current model of exponential growth is unsustainable. We have to come up with alternate models that will address the needs of the poor without destroying the environment.

We will return to this topic after we discuss the real scale of environmental destruction caused by human activities. We will first consider issues such as water scarcity, energy shortage, global warming, climate change, pollution, and the loss of biological diversity. We will then begin to understand why our current practices are unsustainable.

Meanwhile, find out what your ecological footprint is!

Box 3.2 Why We Cling to the Idea of Progress and Unending Economic Growth

- Our culture teaches us that human development is possible only through economic growth. We imbibe this idea from our education, parents, peer group, media, politicians, most economists, and so on.
- We are not aware of other models of development and so we cannot believe that there could be 'development' without constant economic growth.
- We equate making money with happiness.
- We are greatly attracted by the Western way of life that has resulted from economic growth. We yearn for a similar lifestyle and increasing consumption (cars, highways, malls, gadgets, fast food, and so on). We are not aware that the 'happiness index' in countries like the US has been steadily going down.
- Many of us seem to be better off than our parents and grandparents with more comforts and goods. We wish to maintain this trend for ourselves and our children.
- Many of us believe that economic growth will ultimately bring prosperity for all.

END PIECE

The American film director, actor, comedian, writer, musician, and playwright Woody Allen (b. 1935) said:

More than any other time in history, mankind faces a crossroads. One path leads to despair and utter hopelessness. The other, to total extinction. Let us pray we have the wisdom to choose wisely.

EXPLORE FURTHER

Books

Ayres, Ed, 1999, *God's Last Offer: Negotiating for a Sustainable Future*, New York: Four Walls Eight Windows.

CSE, 2000, *Our Ecological Footprint: Think of Your City as an Ecosystem*, New Delhi: Centre for Science and Environment.

Jager, Jill, 2008, *Our Planet: How Much More Can Earth Take?*, London: Haus Publishing.

Leonard, Annie, 2010, *The Story of Stuff: How Our Obsession with Stuff is Trashing the Planet, Our Communities, and Our Health—And a Vision for Change*, New York: Free Press.

McKibben, Bill, 2010, *EAARTH: Making a Life on a Tough New Planet*, Times Books, New York.

Milbrath, Lester W., 1996, *Learning to Think Environmentally While There is Still Time*, Albany, New York: State University of New York Press.

Wackernagel, Mathis and William E. Rees, 1996, *Our Ecological Footprint: Reducing Human Impact on the Earth*, Gabriola Island, Canada: New Society Publishers.

Websites

Ecological Footprint: www.footprintnetwork.org
www.myfootprint.org
www.reducemyecofootprint.com

Films

Films made by Annie Leonard (see website www.storyofstuff.com)

THE CRISIS DESCRIBED

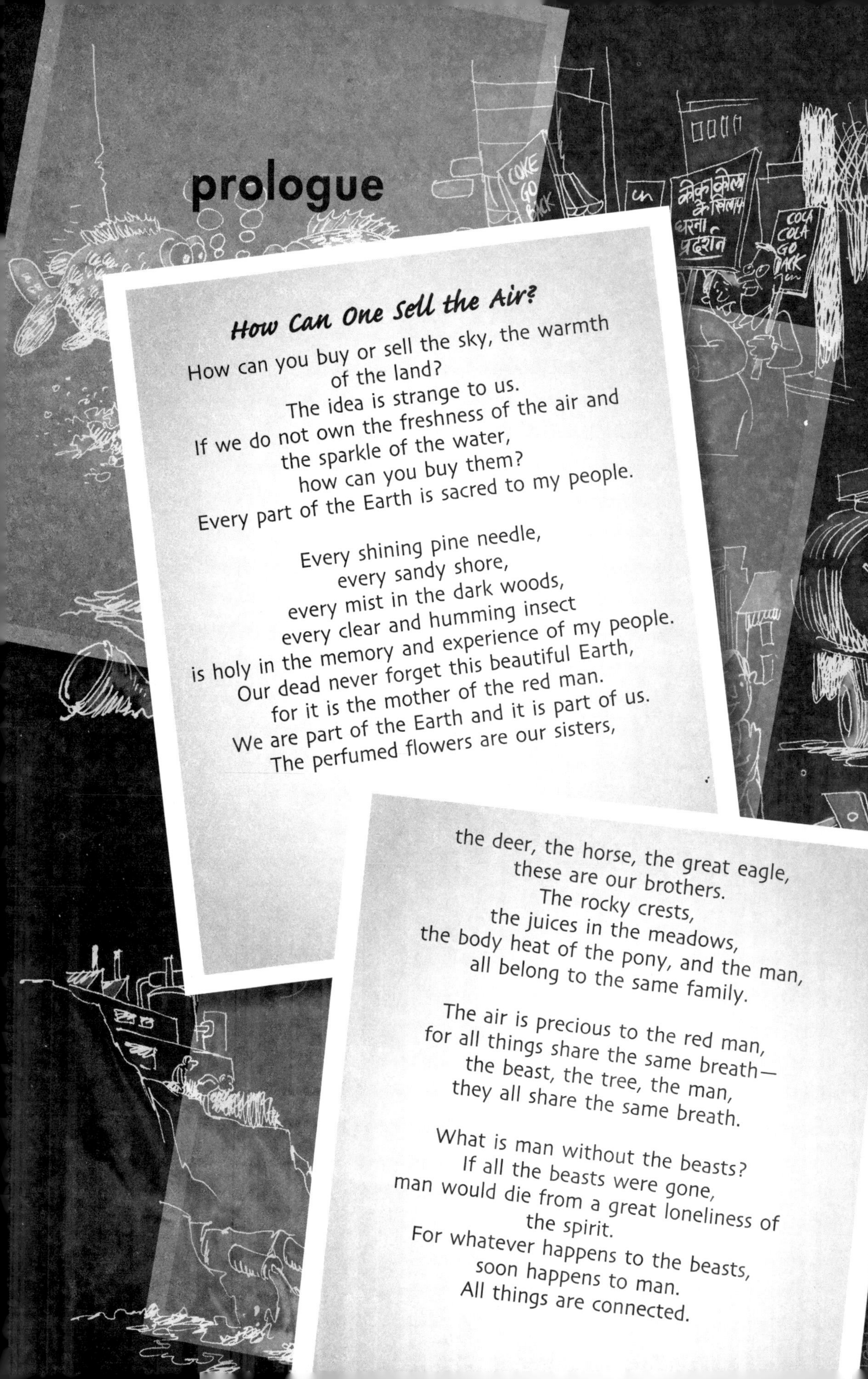

prologue

How Can One Sell the Air?

How can you buy or sell the sky, the warmth
of the land?
The idea is strange to us.
If we do not own the freshness of the air and
the sparkle of the water,
how can you buy them?
Every part of the Earth is sacred to my people.

Every shining pine needle,
every sandy shore,
every mist in the dark woods,
every clear and humming insect
is holy in the memory and experience of my people.
Our dead never forget this beautiful Earth,
for it is the mother of the red man.
We are part of the Earth and it is part of us.
The perfumed flowers are our sisters,

the deer, the horse, the great eagle,
these are our brothers.
The rocky crests,
the juices in the meadows,
the body heat of the pony, and the man,
all belong to the same family.

The air is precious to the red man,
for all things share the same breath—
the beast, the tree, the man,
they all share the same breath.

What is man without the beasts?
If all the beasts were gone,
man would die from a great loneliness of
the spirit.
For whatever happens to the beasts,
soon happens to man.
All things are connected.

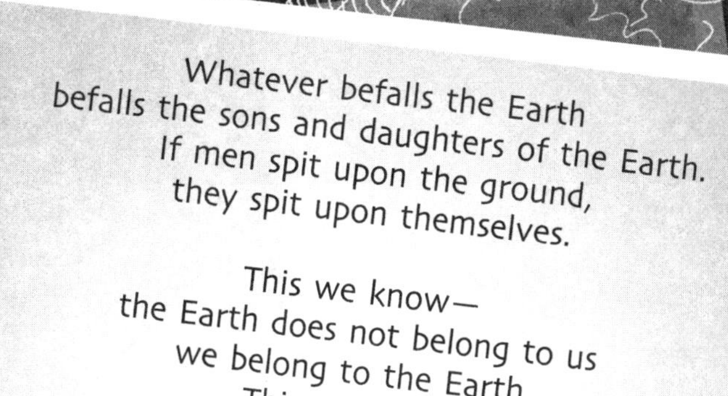

Whatever befalls the Earth
befalls the sons and daughters of the Earth.
If men spit upon the ground,
they spit upon themselves.

This we know—
the Earth does not belong to us
we belong to the Earth.
This we know.
All things are connected
like the blood which unites one family.
All things are connected.

Whatever befalls the Earth
befalls the sons and daughters of the Earth.
We did not weave the web of life
we are merely a strand in it.
Whatever we do to the web,
we do to ourselves.

—Ted Perry

These are excerpts from a speech supposed to have been given in 1854 by Chief Seattle, head of a native American tribe, when a representative of the US government visited him. It was in response to the government's plan to buy the native territory from the tribe. There are different versions of this speech, but the most authentic version is perhaps the one published by Henry Smith in 1887. The most popular version, however, is the one by Ted Perry, a professor at the University of Texas. He wrote it in 1970 as a narration for a film on pollution and ecology and never claimed it as being authentic. The Perry version caught the imagination of environmentalists and activists all over the world and became very popular during the environmental movement of the 1970s. It captures the spirit of modern environmentalism so well that it does not matter that Chief Seattle did not exactly utter those words. The excerpts given above are from the Perry version. The book *Hour Can One Sell the Air? Chief Seattle's Vision* by Gifford and Cook (1992) gives the full text of the different versions and an account of how each one came about.

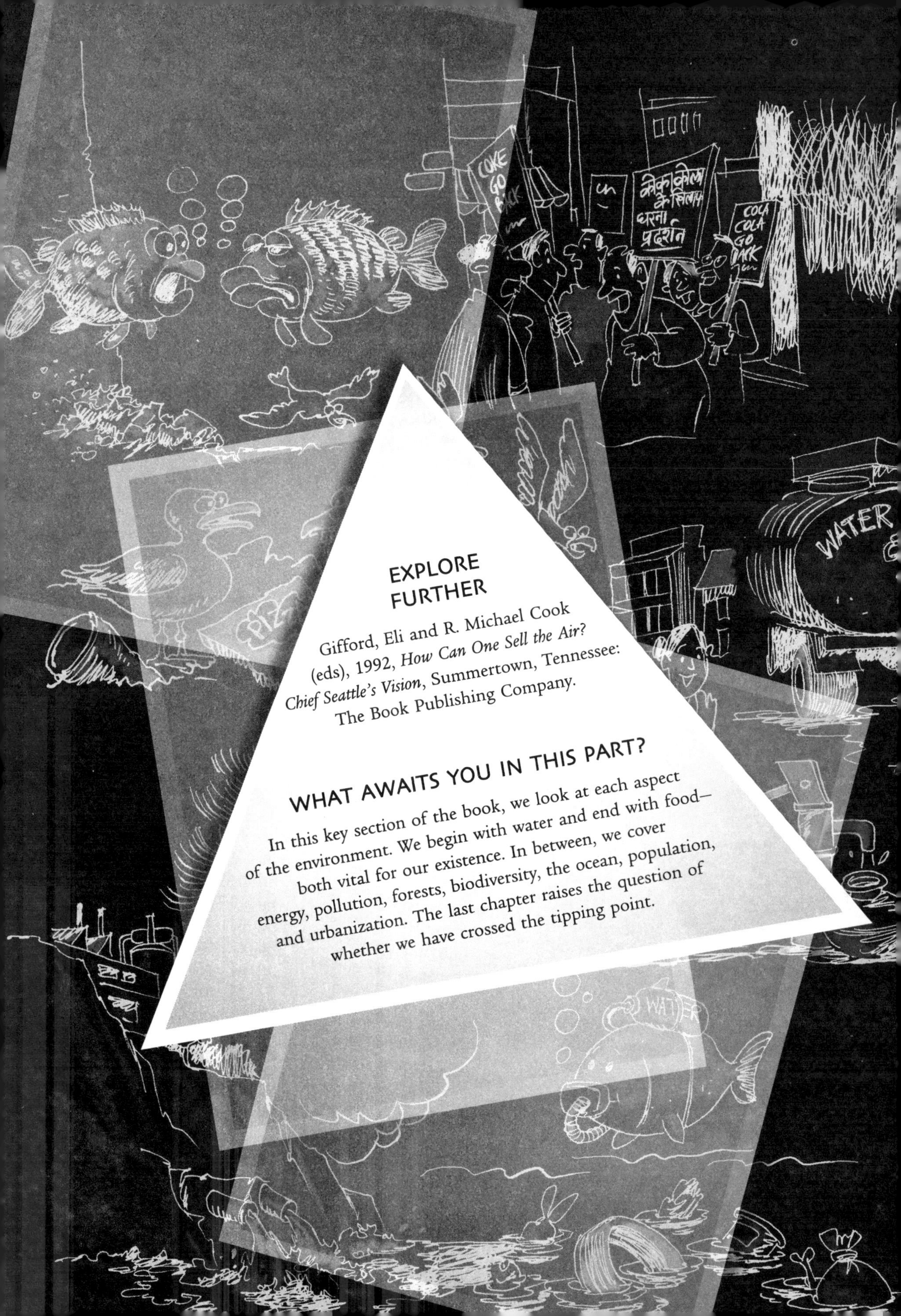

EXPLORE FURTHER

Gifford, Eli and R. Michael Cook (eds), 1992, *How Can One Sell the Air? Chief Seattle's Vision*, Summertown, Tennessee: The Book Publishing Company.

WHAT AWAITS YOU IN THIS PART?

In this key section of the book, we look at each aspect of the environment. We begin with water and end with food—both vital for our existence. In between, we cover energy, pollution, forests, biodiversity, the ocean, population, and urbanization. The last chapter raises the question of whether we have crossed the tipping point.

4 A THIRSTY WORLD Water Scarcity and Water Pollution

The trouble with water is that they are not making more of it.

—MARQ DE VILLIERS*

THE STORY OF CHERRAPUNJI: WETTEST PLACE ON EARTH, NO WATER TO DRINK

When the monsoon comes to Cherrapunji, it does not rain, it pours. It could rain continuously for over two months and you may not see the sun for 20 to 25 days at a stretch. With an average annual rainfall of 11.5 m, it is listed in every geography book as the wettest place on earth.

Some years it rains much more than the average. In 1974, the place received 24.5 m of rain and the record for one day is 1,563 mm. Yet, Cherrapunji, in the state of Meghalaya in north-east India, faces severe drought the rest of the year!

How is it that not a drop of the 11.5 m of rain remains to quench the thirst of the people? The answer lies in the destruction of forests. Once upon a time, the hills around Cherrapunji were full of dense forests. These forests soaked up the heavy rain and released it slowly over the rest of the year. Over the years, however, the forests were cut down. The heavy rains washed away the topsoil, turning the slopes into deserts. It is now a mining town with whole families and immigrants making a living out of the extraction of coal and chalk. The dusty air chokes the life out of people.

* (b. 1940); Canadian journalist and author.

There is no reservoir to store the water. For over twenty years, the residents have depended on piped water supply that comes from far away. That supply is always erratic and undependable. To make matters worse, rainfall has been declining in Cherrapunji. The annual precipitation has come down by 20 per cent over the past five years.

If a place like Cherrapunji suffers from water scarcity, what about the rest of the world?

IS THERE A GLOBAL WATER CRISIS?

We now live in a world of increasing water scarcity.

There is indeed a global water crisis, but it is more pertinent to ask you: Do you have a local crisis? It is very likely that you do not get enough water from the municipal supply and you are buying water. Certainly, you buy packaged water when you travel. In your workplace, water dispensers with bubble cans are common. You are perhaps worried that the water problem is getting worse by the day. You feel that the municipality or the government should do something about it.

In this chapter, we will look at various dimensions of the water crisis and ways of conserving water.

WATER, A PRECIOUS RESOURCE

Water has two qualities that make it a precious natural resource. First, water is absolutely necessary for life. Human beings, animals, and plants cannot survive without water. Second, unlike other natural resources, you cannot (in most cases) replace water with anything else.

The good news about water, however, is that we can reuse it many times. It may change its form, but we can always get it back.

Box 4.1 What You Should Know about Water

- We now live in a world of increasing water scarcity.
- Water is essential to life and is in most cases irreplaceable.
- There is only a finite amount of water in the world that circulates between the land and the ocean.
- We get a regular annual supply of freshwater, thanks to the global water cycle.
- The paradox of the water situation is that there is scarcity amidst plenty, primarily due to the impact of human activities.
- By the middle of the century, most countries in the world (and cities in particular) will be facing water scarcity. India will face severe problems.
- Water scarcity is due to rising population, excessive exploitation of groundwater for irrigation and industry, and the increasing pollution of water bodies due to human activities.
- The world's food security may be threatened due to shortage of irrigation water and there could be future wars over water.
- There are effective ways of conserving and reusing water. Rainwater harvesting and water recycling are two examples.
- There are many examples of successful water conservation efforts in India.
- There are many steps that you can take in your life to conserve water.

MANY NEEDS FOR WATER

Human beings need water for personal use, agriculture, and industries. At home, water is necessary for drinking, cooking food, washing, cleaning, toilet use, and gardening.

In agriculture, we need water for growing food items like rice, wheat, vegetables, fruits, coffee, tea, and sugarcane. Non-food items like cotton also use water. Factories, thermal, and nuclear power plants, and others require large amounts of water. Industries use water for cleaning, as a coolant, and as part of the manufacturing process.

Box 4.2 How Much Water Do You Need Every Day?

In India, the urban water supply systems are designed for a daily consumption of 135 litres per capita (see Table 4.1).

Table 4.1 Average personal water needs of an Indian urban resident

Use	Litres/Person/Day
Drinking	3
Cooking	4
Bathing	20
Flushing	40
Washing clothes	25
Washing utensils	20
Gardening	23
Total	135

Many urban residents consume much more than this norm. (Note how much water the flush toilet consumes!) People in our villages manage with much less water.

Table 4.2 Pattern of Water Use

Sector	Percentage use in the world	Percentage use in industrialized countries	Percentage use in developing countries
Agriculture	70	30	82
Industry	22	59	10
Homes	9	11	8

Table 4.2 shows the rough pattern of water use in the world.

Nature and the rest of the earth need water too! We often forget that animals need water to drink and to clean themselves. A cow, for example, must drink four litres of water to produce one litre of milk. Natural systems like wetlands, lakes, and deltas also need water.

The minimum amount needed for personal use is 20 to 50 litres per person per day, though 100 to 200 litres is often recommended. Adding the needs of agriculture and industry, the minimum total requirement is taken as 1,700 cu. m per person per year. See also the concept of 'water footprint'.

Box 4.3 Virtual Water and Water Footprint

In addition to the direct use of water for our needs, there is also an indirect use of water in making many items that we eat, drink, or use (see Table 4.3). This 'virtual' water is often hidden from us.

Table 4.3 Virtual Water used in Production

Item and Quantity	Water used (litres)
Paper (1 A4 sheet)	10
Potato (100 g.)	25
Cup of coffee (125 ml.)	140
Milk (1 l.)	1,000
Sugar (1 kg.)	1,500
Rice (1 kg.)	3,400
Cotton T-shirt (500 g.)	4,100
Pair of shoes (bovine leather)	8,000
Beef (1 kg.)	16,000

'Water Footprint' is an indicator that measures both direct and indirect water use of a consumer or producer. The water footprint of an individual, community or business is defined as the total volume of freshwater that is used to produce the goods and services consumed by the individual or community or produced by the business. The concept is similar to that of ecological footprint (see Chapter 3).

The water footprint of a country = (Yearly amount of domestic water used to produce all goods and services consumed within the country) + (Yearly amount of virtual water in the goods and services imported into the country). We get the per capita water footprint by dividing the total amount by the population (see Table 4.4).

Table 4.4 Per Capita Water Footprint of Nations

Country	Water Footprint cu.m. per capita per year
US	2,480
Thailand	2,220
Germany	1,545
Pakistan	1,220
India	980
China	700
Global Average	1,240

In terms of total volume of consumption, India has the largest water footprint in the world.

THE EARTH CONTAINS A LOT OF WATER!

The earth holds the same water in the same quantity as it did when it was formed. The earth's water continuously circulates from the ocean to the atmosphere, then to the land and back.

Scientists estimate that the earth contains about 1,400 million cubic km (mn. cu. km). The ocean contains 96.5 per cent of it. Salty groundwater and saltwater lakes on land make up another 1 per cent. The remaining 35 mn. cu. km of freshwater is mostly ice and snow (24 mn. cu. km) and groundwater (11 mn. cu. km).

NATURE GIVES US FRESHWATER EVERY YEAR

The atmospheric water cycle helps us get a regular supply of freshwater. Annually, about 4,30,000 cu. km of water evaporates from the ocean. However, only about 3,90,000 cu. km falls back into the oceans. The remaining 40,000 cu. km moves from the ocean and falls on land as rain and snow.

The water cycle gives us two great benefits. First, we get this supply of water every year. Second, as the water evaporates from the ocean and falls on land, it loses its salt content and thus becomes potable.

SCARCITY AMIDST PLENTY

The annual fresh supply of 40,000 cu. km amounts to about 5,700 cu. m per person per year (for a world population of 7 billion). This appears to be plentiful, since the minimum annual needs of a person is only 1,700 cu. m. Yet, many countries are facing water scarcity!

In 2009, 31 countries (about 480 million people) faced chronic freshwater shortages. By 2025, more than 2.8 billion people in 48 countries (including India) will face severe water problems. By 2050, the number of countries facing water stress or scarcity could rise to 54, with a combined population of 4 billion people–about 40 per cent of the projected global population of 9.4 billion.

Reasons for Water Scarcity

There are several reasons for water scarcity. The distribution of the available water on land is uneven. Some areas of the world get too much water while other places get too little. The transport of water over long distances is impractical.

In many places all the rainfall occurs over a short period in the year leading to floods in monsoon and drought in summer. Aided by deforestation, the water flows quickly into the ocean. Rainfall patterns have also been changing.

The rapidly increasing pollution of rivers, lakes, as well as the groundwater is also reducing the usable supply. Incredibly small amounts of substances like oil can pollute huge amounts of water. Further, many major rivers of the world are drained dry before they reach the sea.

In many places the rates of extraction of groundwater for irrigation are so high that aquifers are getting depleted. As a result, water tables are falling, notably in India, China, and the US.

Water Scarcity in Cities

The megacities of the world, many of them in the poorer countries, need a lot of water and this water is often drawn from the neighbouring villages and far off rivers and lakes. For example, part of the water for Chennai comes from the Krishna River and other places far away. Chennai also draws water from many surrounding agricultural villages. We saw in Chapter 1 how Bangalore gets its water from 95 km away. Apart from increasing the cost of supply of water, such practices also deprive the rural areas of much-needed water.

Even if a city gets good rain, the water is not retained in the ground. Buildings, paving, and roads cover most of the land and the rainwater does not percolate into the ground. Chennai, for example, gets an average rainfall of 1,290 mm per year, which is more than the national average. Yet, 90 per cent of this rain is lost in runoff, evaporation, and flow into the sewage system.

Megacities also pollute the water and release large amounts of wastewater into the rivers and the ocean. Worldwide, less than 20 per cent of urban wastewater is treated.

Leaking taps in cities (and other places) waste enormous amounts of water. In many cities, more than half the available supply is lost through leaks and rotting pipelines.

Water for Irrigation

Today irrigation takes away 70 per cent of the usable water in the world. In many countries, 50 per cent or

more of the food production comes from irrigated land. These include countries like India, China, Egypt, Indonesia, Israel, Japan, Korea, Pakistan, and Peru. By 2025, up to four billion people will live in countries that will lack water to produce their own food.

Even now, inefficient and outdated practices of irrigation continue, leading to wastage of water. Many fields are flooded with water. Though some water gets to the root zone helping the growth of the plant, 50 per cent is wasted as it percolates, evaporates, or runs off.

Most new land brought into agriculture uses groundwater that is depleting fast. Meanwhile, available agricultural land is also lost due to salinization, reservoir siltation, shift of water use, and so on. Salinization is often the result of raising water tables, which happens due to the flooding of the fields. As the water evaporates it leaves behind salt on the topsoil.

The world over, large dams and canal systems are going out of fashion due to spiralling costs, environmental concerns, and displacement of people. Protest movements in many places have stalled the construction of dams.

Globalization and industrialization increases demands for water from industry and urban areas in different parts of the world. The rapid increase in population also means that more water is needed for food and drinking purposes. All these facts seem to indicate that there is not much scope for increasing irrigation at the same rapid rate of the last century.

WATER SITUATION IN INDIA

India and China are counted as water hotspots in the world primarily because of the large population that has to be provided with food and drinking water. Table 4.5 shows how the per capita availability has been going down in India.

In India, more than 60,000 villages are without a single source of drinking water. Diarrhoea, brought on by contaminated water, claims the lives of one million children every year. In addition, 45 million people are affected annually by the poor quality of water.

While the increasing population is one cause of India's water crisis, the problem has been compounded by the steady deterioration, disuse, and disappearance of the traditional tanks and ponds. These water bodies were very effective in retaining the rainwater and recharging the groundwater. The local communities took care of these resources.

Over the last 100 years or more, these tanks were neglected and many of them disappeared. Deforestation of the hills coupled with the absence of tanks has increased the runoff into the sea.

In recent years, the extraction of groundwater using bore wells operated by electric or diesel pumps has reached gigantic proportions. Free electricity and the

Table 4.5 Water Availability in India

Year	Per capita availability of water (cu. m)
1951	5,000
1990	2,400
2007	1,800
2030	1,300 (estimate)

Green Revolution package have sharply increased the water usage. One estimate is that the extraction is already twice the recharge rate. In many states, water tables have been dropping at alarming rates.

As India develops, industries need water more and more. The total available water is limited and hence any increase of supply to irrigation means taking it away from irrigation and domestic use.

Water Pollution and Contamination

Eighty per cent of all illnesses in the poorer countries occurs due to polluted water. The main types water pollutants include organic wastes, infectious microorganisms, fertilizers, pesticides, industrial chemicals, and radioactive substances. Such pollutants adversely affect the health of humans, animals, fishes, and wildlife. The contamination of water bodies is so widespread that no treatment is done in most cases.

Another serious issue is the contamination of groundwater by arsenic and fluoride. Arsenic contamination of groundwater is now a major problem in Bangladesh and West Bengal. Arsenic poisoning first produces skin diseases, leading later to gangrene and cancer. It also brings many other complications like blindness, liver and heart problems, diabetes, and goitre.

Almost 330 million people may be at risk in the two countries. Many future generations are at grave risk from this poisoning. This is perhaps the largest mass poisoning in history, with no easy solution in sight.

Excess of fluoride in water causes fluorosis, which leads to severe bone and joint deformities or to gastro-intestinal and neurological problems. While fluorosis is most severe and widespread in India and China, it is endemic in at least 25 countries across the globe. In India, it is prevalent in 19 states, affecting 65 million people, including 6 million children.

Why has contamination by arsenic and fluoride become a major problem in recent years? Excessive extraction of groundwater has led to deeper and deeper borewells, which draw water from aquifers containing arsenic and high fluoride concentrations.

Box 4.4 Packaged Water: Pure or Poisonous?

In 2002, the Centre for Science and Environment (CSE) in Delhi tested 34 common brands of bottled water for the presence of pesticides. Pesticide residues were found in all the samples except in one imported brand.

The CSE team traced the pesticides to the sources of water, mostly borewells. The bottling plants used a purification process, but it seemed that they could remove only a part of the pesticide residues. In response to the CSE report, the Health Ministry notified new standards for pesticide residues in bottled water.

Thousands of bottles of water are consumed in India every day. If they contain deadly pesticides, why are people not dropping dead on the streets? The reason is that, while the levels may be above norms, they are not so high as to cause immediate symptoms of poisoning. Over a period, however, the pesticides will show their adverse effects on the health.

Here are some facts about the packaged water industry, which is among the fastest growing sectors in the world:

- About 25 to 40 per cent of all bottled water is taken from taps. Yet, the price of bottled water could be as high as 10,000 times that of tap water.
- Three to four litres of water are used for bottling one litre.
- The companies generally get the water at very low cost. The main cost is in packaging and marketing.

However, the environmental and social costs of bottled water include the following:

- Depletion of groundwater.
- Depriving poor villagers of their water.
- Use of petroleum for making the plastic bottles.
- Waste created by used bottles and pouches.
- Transport over long distances using energy and producing emissions.

Box 4.5 Find Out and Be Surprised!

How much water can one drop of oil contaminate, making the water undrinkable?

Answer in *Appendix C*

SOLUTIONS FOR THE WATER CRISIS

Here is a list of possible solutions to the water crisis in India and elsewhere:

1. Reduce demand:
   - Educate people to use less water.
   - Install water-saving devices like self-closing taps and dual-flush toilets.
   - Use decentralized wastewater-recycling systems in homes, apartment blocks, campuses, and industries, using natural methods like planted filters.
   - Adopt composting toilets to save water and also to minimize the sewage disposal problem.
   - Adopt agriculture practices that require less water:
     - Replace water-hungry crops by those requiring less water.
     - Promote crops that can tolerate salty water.
     - Return to indigenous species that can withstand drought.
     - Switch to organic and natural farming.
     - Get more crop per drop—use drip irrigation, sprinkler method, etc.
   - Persuade people to change to a vegetarian diet that would require less water for production.
   - Reduce industrial consumption through recycling, reuse and new water-efficient technologies.
2. Catch the rain where it falls:
   - Retain water on land as long as possible through check dams and contour bunds allowing it to percolate into the ground.
   - Implement rainwater harvesting in urban and rural areas.
   - Restore traditional system of ponds and lakes.
3. Adopt decentralized systems of water supply and sanitation:
   - Plan many small local catchments in place of large ones.
   - Implement a large number of small-scale schemes.
4. Adopt fairer policies:
   - Give control of water sources to the community.
   - Price water properly.
   - Removing inequities in access to water.

COKE GO BACK
कोका कोला के खिलाफ धरना प्रदर्शन
COLA COLA GO
PIZZA
WATER
WATER

PLASTIC
QUEN

Box 4.6 Rainwater Harvesting

There are two ways in which rainwater can be harvested. One way is to slow down the flow of water on land through bunds and check dams. The longer the water remains on land, the more it percolates into the ground. This way it recharges the aquifers and wells. This method is best suited for rural areas.

The second way is to collect and store rainwater. This is a very effective method for our cities. The rainwater that falls on the roof can be collected, filtered, and stored. A surprisingly large amount of water can be collected in this way.

Suppose you live in the city of Delhi. Your house has a terrace area of 100 sq. m. How much of rainwater can you collect in one year?

Average annual rainfall in Delhi = 910 mm
Amount of rain falling on 100 sq. m. area = Roof area x rainfall
= 100 sq. m x 0.91 m
= 91 cu. m
= 91,000 l

For a family of five, consuming 750 l a day, this rainwater will last for 120 days or one-third of the year!

Normally, rainwater is good enough to drink. As a precaution, however, you could avoid using water from the first rain of the monsoon. Rainwater harvesting systems usually incorporate first rain separators. As long as the storage is completely closed, the water remains good for a long period.

Rooftop rainwater can also be used to recharge groundwater. We dig a few percolation pits around the house and fill it with gravel. Water from the roof is directly let into these pits. It percolates into the soil and recharges the groundwater, if the soil is porous. After a while, the water levels in the area will go up and the wells will have enough water. Water from stormwater drains can also be diverted into percolation pits.

At a personal level, you can take several measures to conserve water.

While we can take steps to conserve water, we are unlikely to find dramatic solutions that would once and for all solve the water problem. For example, the plan to interlink Indian rivers will meet insurmountable hurdles such as huge costs, technical problems, ecological impact, and interstate conflicts. There are real limits to the availability of water and we are fast reaching them.

Box 4.7 *What You Can Do to Conserve Water*

Reduce water consumption

- Examine the usage pattern of water in your family or in any one family. Educate them to use water with more care and efficiency.
- Replace leaking taps at home.
- Do not keep the water running while using the washbasin. Replace ordinary taps with self-closing 'railway' taps. Or, remove the taps and keep a bucket of water with a mug.
- Install shower outlet for kitchen sink. It will reduce the amount of water used for cleaning utensils.
- Do the dishes in a bowl rather than under running water.
- Remove any bathtubs you may have.
- Take quick showers instead of baths or take a sponge bath instead of a shower.
- Check the public taps and pipelines in your area, and arrange to plug the leaks.
- Use only low-volume and dual flush toilets.
- Buy a front-loading washing machine. It uses 40 per cent less water than the top-loading one.
- Run dishwashers or washing machines only with full loads.
- Wash your car with a bucket rather than a hosepipe.
- Use drip irrigation for your garden.
- Replace lawns by trees and shrubs that need little fertilizer, and are drought-resistant. Or, grow vegetables.
- Grow only those plants that require low amounts of water.
- Use liberal amounts of mulch to reduce water evaporation.
- Water the garden only after dusk.

Increase supply of water

Implement rainwater harvesting in your house or apartment block.

Bigger Steps

- Recycle bathroom and kitchen water (grey water) using natural methods like planted filters, if you have space.
- Install a composting toilet in your house. This reduces water consumption, reduces the load on the sewage system, and produces organic manure. Collect the urine separately, dilute and apply on plants as fertilizer.

Just to Place Things in Perspective: A cola plant in Madhya Pradesh draws about 1,40,000 cu. m of groundwater per year. This water could have irrigated 2,500 ha of land, sustaining around 5,000 rural families!

Box 4.8 Conflict: Who Owns the Water?

If a soft drink industry owns a piece of land, can it exploit the groundwater without any limit? This is the core issue in what has come to be known as the Plachimada case in Kerala.

In January 2000, Hindustan Coca-Cola Beverages began drawing groundwater and producing cola in Plachimada. The panchayat received an annual income of about Rs 6,00,000 in the deal. In addition, the plant provided employment for about 400 people.

By early 2002, the villagers started noticing changes in the water quality in the area. They felt that the plant was drawing too much water from the ground and this was affecting the harvest of rice and coconuts. In April 2002, the villagers started an agitation against the plant.

The panchayat cancelled the company's licence. The matter went to court, which ordered the plant not to draw water from the ground and to find alternate sources of water. The judge said that the company had no right to extract excessive natural wealth and the panchayat and the government were bound to prevent it. The groundwater was a national resource, which belonged to the entire society and not to the company even though it owned the land.

The Plachimada struggle is still going on and we do not know how the story will ultimately end. The case, however, has become the focal point for the debate on the people's right to water and a company's right to exploit the water under the land that it owns.

Box 4.9 *Hopeful Signs: Water Conservation in India*

In recent times, a number of initiatives have been taken by individuals, communities, NGOs, and even by government agencies to implement measures for water conservation and management. Some of the well-known cases are:

- Auroville: The spectacular efforts of this international community near Puducherry over the past 35 years in afforestation, water conservation, wastewater treatment, and so on, are models to follow. Auroville is now working with local communities and the government to rejuvenate the traditional ponds and get the control back to the people.
- Tarun Bharat Sangh, Alwar, Rajasthan: This NGO, founded by Rajendra Singh, built thousands of earthen check dams in villages with community participation. As a result, water tables rose, dead rivers came back to life, and farming improved.
- Ralegaon Siddhi, Maharashtra: Through the efforts of Anna Hazare, the people have constructed sand percolation tanks and brought back the greenery.
- Gram Gaurav Pratishtan: In Maharashtra, the late Mr Vilas Salunke introduced the idea of a Pani Panchayat for fair sharing of water.
- Tamil Nadu: Rainwater harvesting was made compulsory in the state in 2003.
- Efficient water use by industry: Faced with water scarcity, increasing costs, and protests by local communities, several industries like Arvind Mills, Chennai Petroleum Corporation, and J.K. Papers have implemented water conservation measures and have benefited by them.

Box 4.10 *Water*

If I were called in
To construct a religion
I should make use of water.

Going to church
Would entail a fording
To dry, different clothes;

My liturgy would employ
Images of sousing,
A furious devout drench,

And I should raise in the east
A glass of water
Where any-angled light
Would congregate endlessly.

—Philip Larkin*

* (1922–1985); British poet and novelist.

END PIECE

Here is a quote from Gil Stern, who teaches public speaking in the US:

Man is a complex being; he makes the deserts bloom and lakes die.

EXPLORE FURTHER

Books and Articles

Agarwal, Anil and Sunita Narain (eds), 1997, *Dying Wisdom: Rise, Fall and Potential of India's Traditional Water Harvesting Systems*, New Delhi: Centre for Science and Environment.

Agarwal, Anil, Sunita Narain, and Indira Khurana (eds), 2001, *Making Water Everybody's Business: Practice and Policy of Water Harvesting*, New Delhi: Centre for Science and Environment.

Bhaumik, Subir, 2003, 'World's Wettest Area—Cherrapunji—Dries Up', BBC News (28 April), www.indiaresource.org/news/2003/3923.html

de Villiers, Marq, 1999, *Water*, Toronto: Stoddard.

DTE, 2003, 'Gulp: Bottled Water Has Pesticide Residues', *Down To Earth*, Vol. 11, No. 18 (15 February), pp. 27-34.

Krishna Kumar, R., 2004, 'Resistance in Kerala', *Frontline*, Vol. 21, No. 3 (13 February), pp. 38-40. (Plachimada Case)

Postel, Sandra, 1992, *The Last Oasis: Facing Water Scarcity*, World Watch Environmental Alert Series, London: Earth Scan.

——, 1999, *Pillar of Sand: Can the Irrigation Miracle Last?*, New York: W.W. Norton.

Shresth, Swati with Shridhar Devidas, 2001, *Forest Revival and Water Harvesting: Community Based Conservation at Bhaonta-Kolyala, Rajasthan*, Pune: Kalpavriksh.

Vijayalakshmi, E., 2003, 'Calling the Shots: Cola Major Gets a Taste of Panchayat Power in Kerala', *Down to Earth*, Vol. 2, No. 5 (15 December), pp. 15-19 (Plachimada Case).

Websites

Virtual Water and Water Footprint: www.waterfootprint.org

Films

Drinking the Sky (Joost de Haas, BBC Earth Report, 2000)

5 A WARMING WORLD
Energy, Global Warming, and Climate Change

My father rode a camel.
I drive a car.
My son flies a jet plane.
His son will ride a camel.

—A Saying in Saudi Arabia

THE STORY OF WOMEN HEADLOADERS: BIG BUSINESS, LOW INCOME

Basumati Tirkey, a 35-year-old resident of Bangamunda village in Orissa, walks 9 km every day, carrying a load of 35 kg of fuelwood, to earn Rs 15. And she has spent a lifetime doing so.

Basumati is just one of the 11 million desperate women in India who eke out a living collecting and selling fuel wood. Seventy per cent of women living in and around forest areas are engaged in this work. It is estimated that, in an average village in a semi-arid region, a woman walks more than 1,000 km a year to collect fuel wood alone.

The UN Food and Agricultural Organization (FAO) estimates that the fuel wood business in India has a turnover of USD 60 billion. It must be the most valuable non-timber forest activity and yet it is the last resort of the poorest of the poor. It is generally regarded as being a threat to the forests and an illegal activity. In fact, a substantial part of the huge turnover ends up as payments to the forest guards or profit for the middlemen.

Fuel wood is still the most preferred (and often the only available) energy source in the rural areas and a major source for the urban poor too. It is estimated that 47 per cent of the total energy consumed by households in India is from fuel wood, 17 per cent from animal dung, and 12 per cent from crop residues. This leaves just 24 per cent for commercial energy.

While many millions like Basumati make their living by collecting and selling fuelwood, millions more collect wood just for their daily energy needs. Because of low incomes these poor people are unable to shift to commercial fuels like kerosene.

WHAT DOES THE STORY OF BASUMATI TELL US?

You worry about rising petrol prices and you are in panic when the petrol bunks shut down for a day. Think about the millions of Basumatis in the world, who collect fuel wood every day in the hard way. With increasing population, the demand for fuelwood keeps going up, even as the availability is going down. How will we provide sufficient energy for the poor without degrading the environment?

IS THERE A GLOBAL ENERGY CRISIS?

We now live in a world of increasing shortage and spiralling costs of energy.

Different groups have different perceptions on the question of an energy crisis. This is because of the huge disparity in energy consumption between the rich and the poor. One-third of the world's population, that is, more than two billion people, lack access to adequate energy supplies. At least three billion people depend on fuel wood, dung, coal, charcoal, and kerosene for cooking and heating. On the other hand, the industrialized countries, with only 25 per cent of the global population, account for 70 per cent of the commercial energy consumption.

In countries like the US, any hint of shortage of oil supplies or even a small increase in the price of oil is considered a crisis. In the poorer countries like India, the shortage or increasing price of fuel wood in a village could be a crisis. An urban resident in India faces frequent power cuts, especially in summer, and thus experiences an energy crisis.

THE ENERGY WE NEED

We need energy to drive almost all of our activities:

- Transportation: This is the world's fastest growing form of energy use. That is largely due to the rise of the private car. There are more than 600 million cars on the roads now, and 40-50 million more are added every year. Around the world, we are taking more trips and travelling greater distances.
- Buildings: Energy use in buildings is rising rapidly. The International Energy Agency (IEA) predicts that world electricity demand will double between 2000 and 2030, with most rapid growth in people's homes.
- Manufacturing: We use a major part of global energy for manufacturing vehicles, buildings, appliances, and even food and clothes. We need energy to make an item (embodied energy), to use it during its life (operation and maintenance), and to dispose it off when it is no longer useful (waste management).

GLOBAL ENERGY CONSUMPTION PATTERN

About 24 per cent of energy is used for transportation, 40 per cent for industry, 30 per cent for domestic and commercial purposes, and the remaining 6 per cent for other uses, including agriculture. About 30 per cent of the energy goes into the production of electrical power, which in turn is used by different sectors.

The US is the largest energy consumer in the world. With just 4.6 per cent of the world's population, the US consumes 24 per cent of the total commercial energy produced. India, with 16 per cent of the population, accounts for just 3 per cent of the total energy.

In the US, 92 per cent of the energy used comes from non-renewable fossil fuels that release huge volumes of emissions. The US has 3 per cent of the world's oil, but consumes 26 per cent of the crude oil extracted in the world. Now, many countries, including India and China, are striving to reach the level of industrialization of the US. If we all start consuming energy at the same rate as the US, the world will run out of fossil fuels in just a few years!

SOURCES OF ENERGY

Sources of energy are of two types—non-renewable and renewable. Non-renewable sources are limited in supply and get depleted by use—oil and coal are two examples. Renewables are replenished by natural processes and hence can be used indefinitely—solar energy is an example of this type of energy.

Table 5.1 Share of Different Sources in Total Energy Use

Energy source	Percentage of total energy	Sub-total Percentage
Non-renewable sources		
Oil	32	
Coal	21	
Natural gas	23	
Nuclear	6	
Non-renewables Total		82
Renewable sources		
Biomass (mainly wood)	11	
Solar, wind, hydro and geothermal power	7	
Renewables Total		18
TOTAL	100	100

About 99 per cent of our energy comes from the sun. The commercial energy we pay for is just 1 per cent of the energy we use. Without the sun, life on earth would not exist, since the average temperature would go down to -240°C. It is this solar energy that gets stored in plants as biomass. The plants use the energy for photosynthesis that produces food.

The world's commercial energy comes mostly from fossil fuels like oil, coal and natural gas (see Table 5.1).

Non-renewable Energy Sources

Fossil Fuels

Fossil fuels (coal, oil, and natural gas) are the remains of organisms that lived 200–500 million years ago. During that stage of the earth's evolution, large amounts of dead organic matter had collected. Over millions of years, this matter was buried under layers of sediment and was converted by heat and pressure into coal, oil, and natural gas.

Fossil fuels continue to be formed, but at an extremely slow rate. Today, our consumption rate is far in excess of the rate of formation of fossil fuels. We consume in one day what the earth took one thousand years to form! That is why fossil fuels are considered to be non-renewable.

Worldwide Concern about Oil

The extraction of crude oil from the earth is not a simple process. When a new oil field is tapped, oil gushes out, but not for long. We have to wait for the oil to

Box 5.1 Peak Oil

The production of oil is governed by the Hubbert Curve (see Figure 5.1), proposed by the geophysicist M. King Hubbert. In any oil field, the annual production increases until about 50 per cent of the reserve has been extracted. The production peaks at this point and thereafter declines because it becomes increasingly more difficult and expensive to extract the remaining oil. At some point, the energy expended in extraction exceeds the energy yield from the extracted oil. Then the field is abandoned.

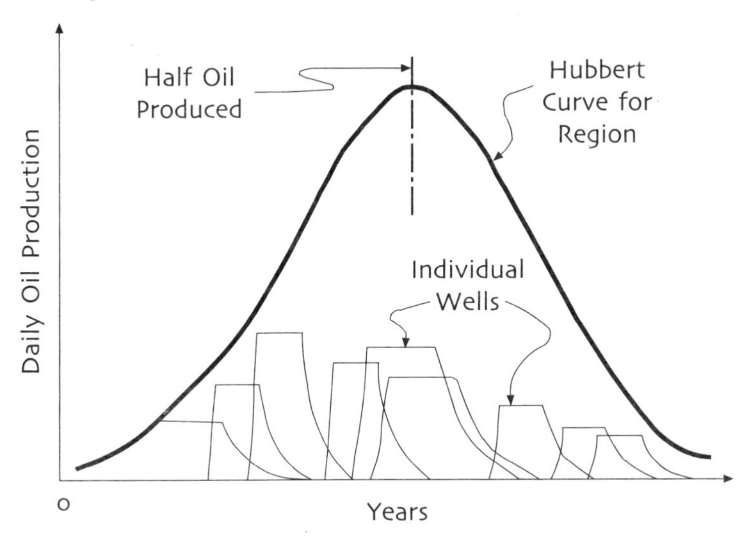

Figure 5.1 The Hubbert Curve for Oil

The Hubbert Curve is applicable also to countries and regions. The production in the US peaked as far back as 1970 and ever since it has been declining. The world peak may have been reached already, or it will do so very soon.

'Oil peak' does not mean 'running out of oil', but it does signify the end of cheap oil. The countries of the Middle East now have most of the oil and, when their production peaks, prices will soar. Any sharp increase in the price of oil will, of course, deeply affect many aspects of our lives.

seep slowly into the well from the surrounding rocks. For this and other reasons, the maximum annual production cannot exceed 10 per cent of the remaining reserves. Oil production is also subject to the phenomenon of peaking.

The environmental costs of oil have been horrendous: air pollution, damage to ecosystems, carbon dioxide emissions, global warming, and so on. Major and minor oil spills have caused untold damage to the ocean, coastal zones, and marine life. In addition, ballast water from the tankers has carried deadly organisms and toxic substances to far-off lands.

How Much Oil is Still Left and How Long Will it Last?

The estimated recoverable reserves are 1.4 to 2.1 trillion barrels. (In the oil industry, the barrel is the preferred unit and it equals 159 litres.) About 70 per cent of the world's crude oil reserves are with the countries of OPEC (Organization of the Petroleum Exporting Countries). In 2010, it had 12 members: Algeria, Angola, Ecuador, Iran, Iraq, Kuwait, Libya, Nigeria, Qatar, Saudi Arabia, the United Arab Emirates, and Venezuela. Saudi Arabia alone accounts for 25 per cent of the total oil reserves.

There are 1,500 major oil fields in operation, of which the 400 large ones account for 60–70 per cent of the production. The discovery of new big fields peaked in 1962 and, since 1980, just 40 have been found. Geologists agree that there are no big ones left for us to discover.

The world demand is about 24 billion barrels per year and is rising. New discoveries, however, amount to just 12 billion barrels per year and this is declining!

Coal

At current rates of use, the world's coal reserves will probably last for another 200 years. However, among the fossil fuels, coal is most harmful to the environment. The mines create major land disturbances and the burning of coal causes severe air pollution. Coal is currently responsible for 36 per cent of carbon dioxide emissions in the world. In addition, it releases huge amounts of radioactive particles into the atmosphere, more than a properly operating nuclear power plant. Every year, air pollution from coal kills thousands of people and causes respiratory diseases in thousands more.

Natural Gas

Natural gas, a mixture of methane, butane, ethane, and propane, is found above most oil reserves. While propane and butane are liquefied and removed as LPG (liquified petroleum gas), methane is cleaned and pumped into pipelines.

About 40 per cent of the total natural gas is in Russia and Kazakhstan. The available reserves are expected to last 200–300 years. However, long pipelines are needed to carry natural gas.

Nuclear Power

In a nuclear reactor, neutrons split the nuclei of elements like uranium and plutonium and in the process release energy as heat. This high temperature heat is used to produce steam, which runs the electric turbine.

When nuclear power first came, it seemed the ideal answer to the energy problem. However, many of the calculations and predictions about nuclear power went wrong. The plants cost much more than the estimates, the operating costs were high, the technical problems were more intractable than expected, and the economic feasibility came to be doubted. The raw material is also in short supply.

Further, the nuclear plants generate large amounts of deadly radioactive waste. The low-level waste must be stored safely for 100–500 years, while the high-level waste remains radioactive for 2,40,000 years! We still have no satisfactory way of storing the waste.

The perceived lack of safety is another issue. Several minor accidents were followed by the major one in Chernobyl, which released radioactive dust over thousands of square kilometres and may ultimately cause between 1,00,000 and 4,75,000 cancer deaths.

Renewable Sources of Energy

Solar Energy

We receive from the sun a pure, non-polluting, and inexhaustible form of energy. Solar energy comes from the thermonuclear fusion reaction constantly taking place in the sun. All the radioactive and polluting by-products of the reaction are safely left behind in the sun, 150 million km away.

An enormous amount of solar energy falls on the earth. What we get from the sun in one month is more than the energy stored in all the fossil fuels we have. Also, using this vast amount of energy does not pollute the biosphere in any way.

Solar energy, however, is a diffused source falling evenly over a vast area. The first problem is to collect it efficiently and the second one is to convert it into a usable form like electricity.

You can guess what the third problem is: What do we do when it is cloudy and the sun does not shine? We must have an efficient way of storing the energy. All research in solar energy is about finding cost-effective ways of collection, conversion, and storage.

Conversion of Solar Energy into Electricity

This conversion is done by a photovoltaic cell (or PV cell), which consists of two layers of silicon. Each cell generates only a small amount of power, but many cells, placed together on a panel, creates enough power to run an appliance.

The power from a solar panel is usually stored in a battery, to which we connect the appliance. Thus the energy is generated and stored when the sun shines on the panel and is used whenever needed. The direct current from the battery can be converted into alternating current through an inverter. You can then run normal appliances like a fan or television.

Solar energy is perennial and non-polluting. However, the capital costs are still high and the efficiency low. Further, there is pollution both in the manufacturing of the panels and disposing them off at the end of their life cycle.

Wind Energy

Wind energy produces electricity at low cost, the capital costs are also moderate, and there are no emissions. Wind farms can be quickly set up and easily expanded.

Obviously, you need steady winds of a certain velocity and not every place is suitable. In any case, you will need some form of backup for windless days. It is a case of high land use, though attempts have been made to use the space below the windmills for agriculture or grazing. There is some noise pollution and the monotonous view of hundreds of windmills is visual pollution. In addition, there is a fear that windmills could interfere with the flight of migratory birds.

Globally, wind farms produce about 25,000 MW of energy. Europe produces 70 per cent of it, with Denmark as the leader. India is now a leading player in the wind energy scene.

Hydropower

Almost 20 per cent of the world's electricity comes from hydropower. In order to get a sizable amount of power, we need a high dam on a river with a large reservoir. The potential energy of the water falling from a height runs the turbine.

The cost of generating hydropower is low and there are no emissions. The reservoir can provide water for irrigation round the year and can also be used for fishing and recreation. It also gives drinking water to towns and cities.

Dams, however, cost a lot of money and take years to build. Most of the suitable rivers of the world have already been dammed and it is now difficult to find new spots. The reservoir drowns large areas of farmland, wildlife habitats, and places of historical and cultural importance. Dams impede the migration of fish along the river and reduce the silt flowing downstream. In fact, sediments pile up against the dam and reduce its useful life.

Large dams also cause large-scale displacement of local communities. The people lose their lands and become environmental refugees. Often, compensation for the lost land is meagre (and not even paid on time) and resettlement never satisfactory. There is a worldwide movement against the building of large dams.

Hydrogen as Fuel

When hydrogen burns and gives energy, it combines with oxygen to produce water vapour. In this process, there is no air pollution or emission of carbon dioxide.

Hydrogen, however, is not available in a free state: it is locked up in water and in compounds like petrol and methane. We need energy and an effective method to get the hydrogen out. Thus the first problem is one of collection.

We have the problem of storage too. Hydrogen is highly explosive and if it is stored as compressed gas, the tank will be large, heavy, and costly. Only large buses and trucks could hold the tanks. If we store it as a liquid, we need very low temperatures and this will require energy.

Fuel Cell

A fuel cell burns hydrogen to produce electricity. In the process, hydrogen combines with oxygen to produce water vapour. Thus there is no pollution and it runs continuously as long as there is input.

Unlike a battery, the fuel cell draws its input (hydrogen and oxygen) from outside. Again, a battery requires recharging, but the fuel cell does not. Finally, there is no toxic output when a fuel cell is discarded.

The development of fuel cells has, however, been slow. There are now experimental buses and cars running on fuel cells, but they are very expensive.

We must find first cost-effective ways of producing hydrogen from water using renewable energy like solar. We should solve all the storage problems too. In the best scenario, we would be using electrical and thermal energy produced by renewable sources like solar and incorporate hydrogen as the fuel for transport.

There are other renewable sources like tidal energy, ocean thermal energy, geothermal energy, and energy from biomass (plant material and animal wastes). They are useful as small-scale alternatives, but cannot yet satisfy the world's enormous appetite for energy. In particular, they are not yet suitable for transportation.

ENERGY SCENE IN INDIA

In India, about one-third of the energy comes from non-commercial sources. The rural population depends heavily on fuel wood, dung, and animal waste. In the urban areas, there are large numbers of non-motorized vehicles like bicycles, rickshaws, handcarts, and animal carts.

India ranks sixth in the world in total energy consumption. Our per capita energy consumption is 290 units (kg of oil equivalent), compared to 8,000 units in the US and 600 units in China. Though we have abundant coal reserves, we have very little oil. More than 25 per cent of our primary energy needs is met by import of crude oil and natural gas.

Of the commercial energy used in India, 51 per cent comes from coal, 36 per cent from oil, 9 per cent from natural gas, 2 per cent from hydropower, and just 2 per cent from nuclear plants.

India produces about 33 million tons of oil and 32 million tons of natural gas. We import more than 90 million tons of oil, that is, over 70 per cent of our oil needs. The oil import bill is about USD 90 billion.

We have major programmes for renewable energy. The government is promoting wind farms, solar energy, hydropower, as well as waste-to-energy projects. Several wind farms have been set up in south India. Solar pumps and water-heaters can be found in many parts of the country.

Box 5.2 What You Should Know about Energy and Climate Change

- We now live in a world of increasing energy shortage and higher costs.
- Nearly 75 per cent of the commercial energy in the world comes from non-renewable fossil fuels (coal, oil, and natural gas).
- The rate of consumption of fossil fuels, especially oil, is far in excess of their rate of formation.
- Excessive burning of fossil fuels releases huge amounts of greenhouse gases, which leads to global warming. This in turn results in massive changes in climate, natural disasters, biodiversity loss, and sea level rise.
- The demand for energy in India is racing ahead of supply. The country depends increasingly on import of oil.
- There will be serious environmental and social consequences, if India and China seek to copy the energy consumption levels of the richer countries.
- Energy efficiency can and must be increased.
- Almost all renewable sources of energy have their limits and problems.
- We are unlikely to find a clean, simple, and inexhaustible source of energy. The only way is to reduce our consumption levels.

CONSERVATION OF ENERGY

It is clear that we must conserve energy by improving efficiency at each of the stages: production, transmission, and utilization. According to estimates, we can save up to 40 per cent of energy in India through conservation. The Energy Conservation Act (2001) specifies energy standards and promotes energy audits.

The Bureau of Energy Efficiency (BEE) is the agency that implements the Act. Among other activities, BEE awards star ratings to brands of appliances such as tubelights, refrigerators, and air-conditioners. Each tested brand gets a rating of 1 star (least energy efficient) to 5 stars (most energy efficient). The voluntary scheme will be made compulsory later.

WHAT DOES IT ALL AMOUNT TO?

No fuel is going to save us if we keep increasing our energy use or even maintain the current usage levels. Once again we face the same hurdle: There is a natural limit to the amount of resources on this planet. When we overstep that limit, we will inevitably face an insurmountable hurdle. We can survive only if we use energy much more efficiently and reduce our consumption levels. As of now, there is no sign of this happening.

Let us now turn to global warming and climate change, which are intimately connected with our use of energy.

GREENHOUSE GASES AND THE GREENHOUSE EFFECT

While in school, you would have learned how the carbon cycle works: during photosynthesis, plants absorb carbon dioxide and release oxygen. Organisms breathe in this oxygen, and give out carbon dioxide, which goes back to the plants.

Normally, carbon dioxide and other gases that surround the planet let the radiation from the sun reach the earth. They, however, prevent some of the heat from being reflected back. Since they trap the sun's radiation somewhat like glass does in a greenhouse, we call them greenhouse gases. The resultant warming of the atmosphere is called the greenhouse effect. Without these greenhouse gases, the earth would be far colder, largely covered with ice.

Human activities, however, have been increasing the greenhouse effect. By burning large amounts of fossil fuels, we release huge quantities of carbon dioxide into the atmosphere. Concurrently, deforestation also releases carbon trapped in the tissues of the trees. At the same time, loss of trees reduces the earth's capacity to absorb carbon dioxide through photosynthesis.

We saw in Chapter 3 how the carbon dioxide concentration in the atmosphere shows an exponential growth. Due to the factors listed above, the concentration has been rapidly increasing in the recent years. This is shown by the famous Keeling Graph (Figure 5.2). The safe limit is 350 ppm, but we crossed it many years ago.

Most of the greenhouse gas emissions come from the northern hemisphere, the US being the highest contributor. China and India are industrializing rapidly and their emissions are likely to double over the next two decades.

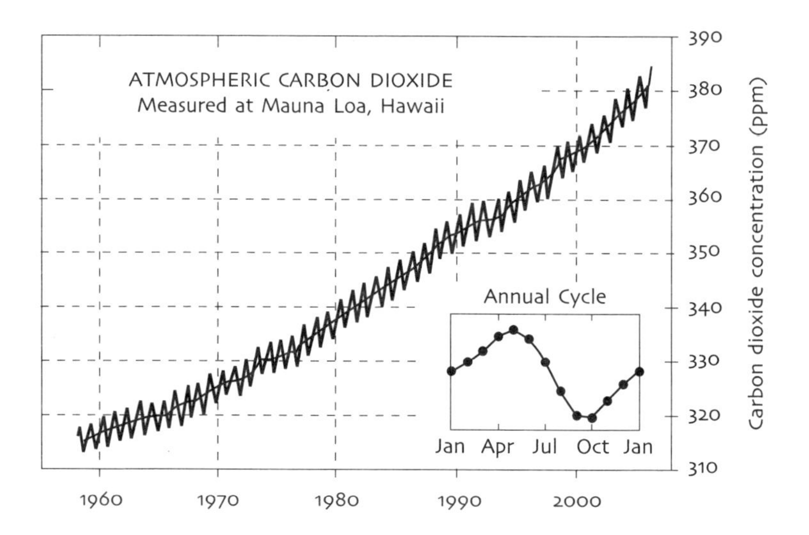

Figure 5.2 Keeling Graph

GLOBAL WARMING

The abnormal increase in the concentration of the greenhouse gases is resulting in higher temperatures. We call this effect as global warming. The average temperature around the world has increased about 1°C over 140 years, 75 per cent of this has risen just over the past 30 years.

Why should we worry about a rise of just one degree? Even a small change in the global temperature can have major consequences. About 20,000 years ago, when ice covered most of Europe, the mean temperature was only five degrees colder than today. Over the entire period of human existence, the Earth has never been more than a degree or two warmer than it is today.

Further, the one-degree rise is in the mean temperature. The northern hemisphere is warming faster than the southern half of the planet. Dramatic melting of ice and snow is occurring at the poles and mountainous areas.

Effects of Global Warming

Accurate predictions are difficult, but all computer models indicate an average rise of 3°C by 2100. An increase of just 1.5°C in the mean global temperature could cause a major change in the climate. The change is likely to be greater than anything experienced during the last 10,000 years.

Global warming is likely to have a wide variety of effects on the following:

- Climate change.
- Ocean and coasts.
- Glaciers, ice caps, and permafrost.
- Water, agriculture, and food.
- Animal and plant species.

These effects are interconnected. In fact, climate change causes many of the other effects.

Climate Change

Regional and seasonal weather patterns will change, with longer summers and shorter winters. Extreme weather conditions like floods and droughts are likely to occur more often. Across the world the effects may be even contradictory, due to the interconnectedness of the ocean, land, and atmosphere. For example, when one region experiences floods, another region may suffer severe drought.

Many scientists have pointed towards some climate change indicators that are already noticeable. Since the 1960s, each decade has been warmer than the previous one. In the 1990s, there were an unprecedented number of natural disasters. During that decade, the weather-related damage was five times greater than in the 1980s. More recently, the years 2002 to 2005 were the warmest on record.

Ocean and Coasts

The ocean has become warmer and sea levels are rising. The melting of polar ice caps is adding to the problem. Small islands like those of the Maldives and Tuvalu are threatened. In fact, Tuvalu may be the first country to go off the world map due to climate change. Two islands in the Sunderbans have already disappeared. If the rise in sea level continues, coastal areas will be flooded in places like the Netherlands, Egypt, Bangladesh, and Indonesia, necessitating the evacuation of large populations.

Melting of Glaciers, Ice Caps, and Permafrost

The most dramatic evidence of global warming is the melting of many glaciers and the Arctic Ice.

Water, Agriculture, and Food

Extreme floods and droughts are likely to have serious effects on water resources, agriculture, and food security. Some recent events that have been attributed to global warming include:

DAM!
WATER
POWER PLANT

CARBON
FOOTPRINT
WARNING
TURN OFF
ENGINE WHILE
FUELING.
2

- Severe water scarcity in Australia that affected irrigation.
- The Yellow River in China, which is a water source for millions, being threatened by shrinking glaciers and lakes.
- The melting of the world's third largest ice field in Nepal, threatening the flood densely populated valley below.

Animals, Plants, and Human Beings

Thousands of animal and plant species will go extinct, unable to adjust quickly enough to the new conditions. Polar species may be the first to go, followed by those in the coastal zones everywhere. Polar bears, which depend on ice, are in big trouble and could be extinct by 2050. Animals and birds will change their migration patterns.

Box 5.3 Why We Disagree about Climate Change

Why do we disagree about climate change, especially about the action to be taken to mitigate its impact? Why are individuals, communities and governments unwilling to commit themselves to action?

Alarmist messages of impending chaos and catastrophe due to climate change seem to be counter-productive. They only generate a feeling of helplessness and apathy. They do not lead to behavioural change among the citizens.

Some of the reasons why we (the people and governments of the world) disagree on climate change are:

- We receive multiple and conflicting messages about climate change and interpret them in different ways.
- We understand science and scientific knowledge in different ways.
- Our political ideologies and our 'development' priorities are different.
- We value things (activities, people, assets, resources) differently.
- We believe different things about ourselves, the universe, and our place in the universe.

Perhaps we should respond to climate change or the other major environmental problems not by trying to 'solve' them, but in creative ways. For example, we should examine closely the long-term implications of our short-term actions. We could also ask what we really want for ourselves and humanity.

Warming will increase photosynthesis activity leading to faster growth of plants and trees. Initially, the yields will be more, but too much heat will kill the crops.

Extreme weather will increase human migration. There will be many millions more of environmental refugees. People living on the coasts will suffer extensive damage due to sea-level rise and cyclones.

Impact of Global Warming on India

According to a report issued by the Indian Meteorological Department (IMD), 2010 was the warmest year in India since 1901. Further, of the 10 warmest years since 1901, seven were recorded since 2002. The average annual temperatures in 2010 were above normal by one degree throughout the country. The average temperature increased by 0.5°C since 1901, most of it happening since 1990 (in tune with global trends).

Global warming has already affected our monsoon patterns and climate. Since 1951, instances of very heavy rains every year have gone up. Spells of moderate rain are going down. This changing trend of the monsoon has made us more vulnerable to disasters such as floods and landslides.

Frequencies and intensities of tropical cyclones in the Bay of Bengal could increase, particularly in the post-monsoon period. Low-lying coastal areas could be flooded, displacing millions of people. Mumbai and other coastal cities could suffer severe losses.

Some states are likely get very heavy rains, while other states could experience severe droughts. The meltdown of Himalayan glaciers would ultimately lead to water scarcity and severe fall in food production.

Responses to Global Warming

By the early 1990s, it was clear to many scientists that global warming was a serious issue. However, the petroleum and automobile industries tried to cast doubts over global warming. They were against any move to cut carbon emissions. They did not want any steps that would affect their businesses and profits. The US government too supported them. The US did not want its economy to be hit, or the citizens' lifestyles to be changed.

Meanwhile, the UN had set up (in 1988) the Intergovernmental Panel on Climate Change (IPCC), a task force of climate scientists from nearly 100 countries. The IPCC issues regular reports on climate change.

Over the years, the IPCC reports have become increasingly strong in stressing the seriousness of global warming and the need for action. The IPCC report of February 2007 said that global temperatures could rise up to 6°C by 2100, triggering disaster for billions of people. In its May 2007 report, the IPCC stated that greenhouse emissions must start declining by 2015, if we wanted to prevent temperatures from rising more than 2°C over the pre-industrialization levels.

Kyoto Protocol

As the IPCC began issuing its reports from 1990, the UN initiated international negotiations on steps to cope with and contain global warming. Many meetings followed and ultimately, the Kyoto Protocol was approved in 1997. It was a legally binding international agreement to reduce greenhouse gas emissions.

The protocol committed the industrialized countries to reducing emissions of six greenhouse gases by 5 per cent by 2012. The target for the US was 7 per cent. Developing countries were not bound to emissions reduction targets.

Box 5.4 *The Story of the Arctic: Melting Ice, Threatened Bears*

The Arctic, the area around the North Pole, is a vast ocean covered with ice and surrounded by treeless, frozen ground. The Arctic region is one of the last remaining areas of unspoilt nature.

The Arctic is full of organisms living in the ice: fish and marine mammals, birds, and land animals, besides human settlements. Its biological diversity is very valuable to us. The natural resources include oil, gas, minerals, forests, and fish.

The Arctic region is very sensitive to changes in the climate. Thus, it acts as an early warning system for the earth. In fact, it is now giving us a major danger signal: the Arctic ice is melting incredibly fast! The reason is global warming caused by human activities.

The North Pole has been frozen for 1,00,000 years. That will, however, not be true by the end of the twenty-first century! During the past 15 years, more than 4,00,000 sq. km of ice has melted away. Greenland has been losing 50 cu. km of water per year from its vast ice sheet.

Summer temperatures are higher than ever. There is a consensus among the scientists that, by 2030, the Arctic will be free of ice in summertime. As snow and ice melt, they expose land and water that absorb more solar heat. That melts more snow and ice, and so it goes on.

Arctic snow and ice play a key role in controlling the planet's temperature. They act as insulation, keeping heat and moisture in the land and ocean and out of the atmosphere. But once the ice and snow are gone, the climate will change in ways that we cannot even predict.

Warming of the ocean and melting of ice will lead to sea level rise, which will submerge small islands and cause enormous damage to many coastal areas. Scientists also predict severe storms and hurricanes.

cont'd ...

Even if the Kyoto Pact had been fully implemented, it would have done very little to solve the problem of carbon dioxide emissions. The IPCC had stated in its 1995 report that the emissions had to be cut by 60–80 per cent if the climate had to be stabilized.

Many countries including India ratified the protocol and it came into force in 2005. The US, however, did not sign the Protocol. A meeting was held in Copenhagen in December 2009 to work out a new protocol. But no effective agreement could be reached.

... cont'd

There is another big danger: Huge permafrost area in western Siberia, the size of France and Germany combined, has begun to melt for the first time since it was formed at the end of the last ice age. It is now releasing methane. Frozen organic matter in the permafrost holds one-seventh of the world's carbon. Its release could dramatically increase carbon dioxide concentration in the atmosphere. This will worsen global warming.

The polar bears and other Arctic creatures, including birds, are in trouble. Because the summers are longer and there is generally less ice, their movements and feeding patterns are getting disrupted. Many polar bears drown since there is more water and they cannot swim for long. Polar bears have been declared a threatened species.

The Arctic region is sparsely populated and the people who live there, especially the native residents like the Inuit, do not have lifestyles of high consumption. Yet, they have been among the first to experience the impact of global warming caused by the energy consumption of people in other parts of the world.

Taking Action on Global Warming

On the whole, a grim future awaits the earth if the emissions continue at the current rates. Even if the greenhouse emissions were to be drastically decreased now, the effects of the past emissions will continue for a long time. Yet, countries are unwilling to sacrifice some comforts today to avoid catastrophes tomorrow.

It is unlikely that global warming can be avoided, but we can reduce its adverse effects. Greenhouse gas emissions must be reduced immediately and drastically as recommended by the IPCC. Every organization, company, and country must measure and reduce its carbon footprint.

Reduction of energy consumption, promotion of renewable energy sources, production of cleaner and fewer automobiles, greater support for public transportation, reduced deforestation, invention of cleaner technologies, and similar measures are urgently needed.

Box 5.5 Carbon Footprint

In addition to ecological footprint and water footprint, we also have the concept of carbon footprint. It measures the total greenhouse gas emissions caused directly and indirectly by a person, organization, event, product, or country.

Unlike ecological footprint, carbon footprint is not measured in area, but in tonnes of carbon-dioxide equivalent (tCO_2e). The footprint of a country does not just measure emissions that occur on its territory.

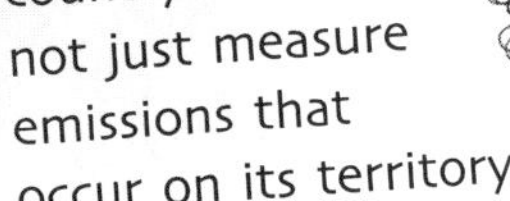

Similar to water footprint, it includes emissions that occur in the production of all goods and services consumed in a country. It also includes international transport (ocean freight and aviation).

Table 5.2 gives the carbon footprint of selected countries.

Table 5.2 Carbon Footprint of Nations

Country	Carbon Footprint (tCO_2e)
US	28.6
Singapore	24.1
Canada	19.6
Germany	15.1
Japan	13.8
South Africa	6.0
China	3.1
India	1.8
Bangladesh	1.1
Global Average	5.1

What You Can Do to Conserve Energy

Each one of us can take action to conserve energy in our personal life and work places. We can save energy at home, in the office, and during transportation.

Just to Emphasize the Point: During the time it took you to read this chapter, the world has consumed about one million barrels of oil!

Box 5.6 *Hopeful Signs: National Action Plan on Climate Change*

On 30 June 2008, Prime Minister Manmohan Singh released India's first National Action Plan on Climate Change (NAPCC) outlining existing and future policies and programs addressing climate mitigation and adaptation. He had earlier set up the Prime Minister's Council on Climate Change (PMCCC).

The NAPCC identifies eight core 'national missions', including:

- National Solar Mission: To promote the development and use of solar energy for power generation and other uses with the ultimate objective of making solar competitive with fossil-based energy options.
- National Mission for Enhanced Energy Efficiency: To promote energy consumption decreases in large industries and offer energy incentives.
- National Mission on Strategic Knowledge for Climate Change: To gain a better understanding of climate science, impacts and challenges, improve climate modelling, and develop adaptation and mitigation technologies.

The NAPCC also includes the ongoing initiatives:

- Power Generation: The government is mandating the retirement of inefficient coal-fired power plants.
- Renewable Energy: The central and the state electricity regulatory commissions must purchase a certain percentage of power from renewable sources.
- Energy Efficiency: Large industries are required to undertake energy audits.

Box 5.7 *Find Out and Be Surprised!*

1. Until 1958, about 115 billion barrels of oil had been drilled out of the earth. Can you guess how long it took to extract the next 115 barrels of oil?

2. If a Chinese citizen consumed oil in amounts equal to an average American citizen, how much oil would China need every day?

Answers in *Appendix C*

Box 5.8 *Save Electricity at Home and in the Office*

- Turn off lights and fans when you leave a room.
- Shut off personal computers, television sets, set-top boxes, music systems, and others, when not in use. All appliances consume energy even on standby mode.
- Install automatic switch-off devices for areas like staircases.
- Replace all the bulbs in your home with Compact Fluorescent Lamps (CFLs), or even better, with Light Emitting Diodes (LEDs). Thanks to lower power bills, you can recover the cost in a few months.
- If you can afford it, install a solar lighting system. Apart from some saving electricity, it will give you light during power breakdowns.
- If you use hot water, install a solar water heater in place of an electric geyser.
- Use a solar cooker, say, for rice and *dhal*. It saves gas and the food tastes better.
- Buy energy-efficient appliances when you replace old ones or you need new ones. Check always the specifications for energy consumption figures and look for the BEE star rating.
- Iron a pile of clothes at a time, instead of one or two at a time.
- Run a washing machine only when there is a full load.
- Avoid using kitchen machines every day. Grind spices once or twice a week.
- Build your house following the principles of ecological architecture.

Box 5.9 *Save Energy in Transportation*

- Minimize the use of automobiles for your personal transport.
- Keep your vehicles tuned for low consumption of fuel.
- Avoid idling your vehicle at signals and traffic jams. Ten seconds of idling uses more fuel than in restarting the vehicle.
- Use bicycle for local work like shopping.
- Use public transport whenever possible. One busload of people takes 40 vehicles off the road during rush hour, saves 70,000 litres of petrol, and avoids over 175 tons of emissions every year.
- Join car pools.
- Live near your place of study or work, if possible.
- Check fuel consumption data while buying a new vehicle.
- Avoid air travel; if you can, go by train instead.

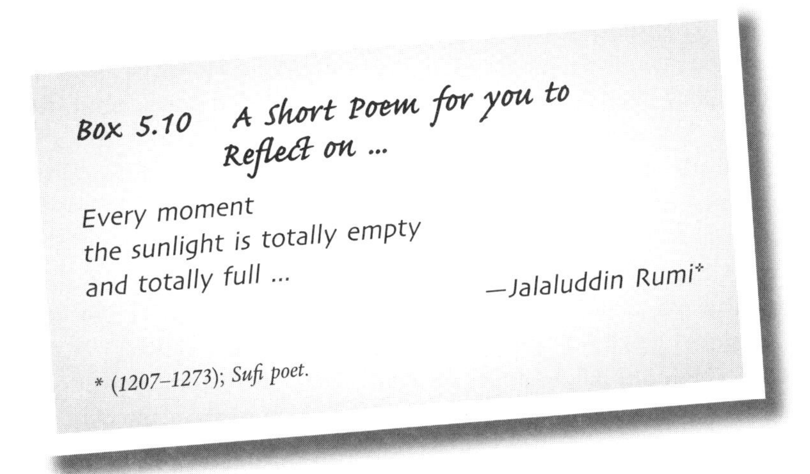

END PIECE

Being a supporter of big businesses, US President George Bush did not want to take any action to reduce emissions and combat climate change. In fact, he first denied the very occurrence of global warming. This led the comedian and TV host Jay Leno to say:

According to a survey in this week's Time *magazine, 85 per cent of Americans think global warming is happening. The other 15 per cent work for the White House.*

EXPLORE FURTHER

Books and Articles

Anderson, Alun, 2009, *After the Ice: Life, Death, and Geopolitics in the New Arctic*, Washington, D.C.: Smithsonian.

Craven, Greg, 2009, *What is the Worst that Could Happen? A Rational Response to the Climate Change Debate*, New York: Perigee Books, Penguin Group.

Das, Anjana and Vikram Dayal, 1998, 'An Energy Shortage Closer to Home', *The Economic Times* (15 January) (Fuel wood crisis/women headloaders).

Dauncey, Guy, 2009, *The Climate Challenge: 101 Solutions to Global Warming*, Gabriola Island, Canada: New Society Publishers.

Gelbspan, Ross, 2005, *Boiling Point: How Politicians, Big Oil and Coal, Journalists and Activists Are Fueling the Climate Crisis—And What We Can Do to Avert Disaster*, New York: Basic Books.

Hansen, James, 2009, *Storms of My Grandchildren: The Truth About the Coming Climate Catastrophe and Our Last Chance to Save Humanity*, New York: Bloomsbury.

Homer-Dixon, Thomas, 2009, *Carbon Shift: How the Twin Crises of Oil Depletion and Climate Change Will Define the Future*, Toronto: Random House.
Hulme, Mike, 2009, *Why We Disagree About Climate Change: Understanding Controversy, Inaction and Opportunity,* Cambridge, UK: Cambridge University Press.
Kolbert, Elizabeth, 2007, *Field Notes from a Catastrophe: Man, Nature, and Climate Change*, London: Bloomsbury.
Lynas, Mark, 2008, *Six Degrees: Our Future in a Hotter Planet*, Washington, D.C.: National Geographic Society.
Mahapatra, Richard, 2002, 'Phulmai's Walk', *Down To Earth,* Vol. 11, No. 14 (15 December), pp. 25–34 (Fuel wood crisis/women headloaders).
———, 2003, 'The Multi-billion-dollar Fuelwood Trade is the Last Resort for India's Poor', InfoChange News & Features, www.infochangeindia.org/features112.jsp).
Monbiot, George, 2007, *Heat: How to Stop the Planet From Burning,* London: Penguin Books.
Orr, David W., 2009, *Down to the Wire: Confronting Climate Change*, New York: Oxford University Press.
Rubin, Jeff, 2009, *Why Your World is About to Get a Whole Lot Smaller*, Toronto: Random House.

Websites

Arctic: http://nsidc.org/arcticseaicenews/
Arctic: www.arctic.noaa.gov/detect/index.shtml
Carbon footprint of nations http://pubs.acs.org/doi/abs/10.1021/es803496a
Climate change calculator: www.nature.org/initiatives/climatechange/calculator/
Climate change: www.climatechange.net
Energy: www.energyquest.ca.gov/story/index.html
Energy: www.planetforlife.com
Fuel wood crisis: www.handels.gu.se/econ/EEU/
India's National Action Plan (NAP) on Climate Change: http://pmindia.nic.in/climate_change.htm
Siberian permafrost: www.newscientist.com, www.bbc.co.uk
The Energy and Resources Institute (TERI), New Delhi: www.teriin.org
UN Framework Convention on Climate Change (UNFCCC): www.unfccc.int
UN Intergovernmental Panel on Climate Change (IPCC): www.ipcc.org

Films

An Inconvenient Truth (Al Gore's documentary on global warming, Davis Guggenheim, 2006)
A Crude Awakening: The Oil Crash (Basil Gelpke and Ray McCormack, 2006)
Age of Stupid (on global warming, Franny Armstrong, 2009)

6 A TOXIC WORLD
Waste, Pollution, and Environmental Health

And Man created the plastic bag and the tin and aluminium can
and the cellophane wrapper and the paper plate, and
this was good because Man could then take his automobile and
buy all his food in one place and
He could save that which was good to eat in the refrigerator and
throw away that which had no further use.
And soon the earth was covered with plastic bags and aluminium cans and paper plates and disposable bottles and
there was nowhere to sit down or walk, and
Man shook his head and cried: 'Look at this Godawful mess.'

—ART BUCHWALD*

THE STORY OF BHOPAL GAS TRAGEDY: THE LUCKY ONES DIED THAT NIGHT

It was five minutes past midnight in Bhopal on 2 December 1984. The congested city was asleep and the winter air was heavy. Many had gone to sleep late after watching the Sunday movie on television. It was all quiet, but the city was to change forever.

*(1925–2007); American humourist and author.

Suddenly, 27 tons of lethal gases including methyl isocyanate (MIC) started leaking from Union Carbide's pesticide factory. The deadly cloud of gases rapidly blanketed the city. It turned out to be the world's worst industrial disaster.

The gases affected more than 5,00,000 people, leaving them with a lifetime of ill-health and mental trauma. More than 8,000 people died in the few days following the disaster. Almost 70,000 people were evacuated from the area after the accident and 2,00,000 more fled in panic. The effect of the gas was felt up to 100 sq. km from the factory. Most of the victims were poor people living in the slums that surrounded the factory.

Today, at least 1,50,000 people, including children born to gas exposed parents, continue to suffer health problems such as headaches, breathlessness, giddiness, numb limbs, body aches, fevers, nausea, anxiety attacks, neurological damage, cancers, depression, and mental illness. The death toll rose to more than 20,000, and at least 30 people continue to die every month due to exposure related illnesses.

How did the accident occur? Most probably water entered the storage tank and caused a runaway chemical reaction that led to an increase in temperature, which converted the liquid MIC into gas. The safety systems did not work.

Union Carbide accepted only moral responsibility for the disaster and not any liability. The Government of India filed a compensation case against the company for USD 3 billion, but strangely accepted USD 470 million as settlement in 1989. Nearly 95 per cent of the survivors have received just Rs 25,000 for lifelong injury and loss of livelihood. That works out to less than Rs 4 a day for more than 20 years of unimaginable suffering.

The story did not end with the disaster. When Union Carbide finally quit Bhopal in 1998, it left behind around 5,000 tons of deadly waste. The toxins have since leached into the soil and water in and around the factory. More than 20,000 local people, 70 per cent of them gas affected, are facing soil and water contamination. Dow Chemical, which purchased Union Carbide, has refused to clean up the site, provide safe drinking water, or compensate the victims.

Even the small amount given by Union Carbide to the Indian government was not fully disbursed. In 2004, the Supreme Court of India ordered the government to distribute the balance of Rs 15 billion in the compensation fund among the 5,72,000 victims.

For over 25 years, survivors' organizations in Bhopal have refused to give up their fight for justice, proper compensation, economic rehabilitation, and adequate health care. Their struggle goes on.

WHAT ARE THE LESSONS FROM THE BHOPAL TRAGEDY?

The tragedy showed that poor communities are disproportionately affected when toxic materials are discharged into the air, land. and water. When a crisis occurs or an accident happens, these people cannot easily get justice from the polluters or from the governments. The poor often do not know how dangerous their workplaces or neighbourhoods are.

WASTE AND POLLUTION

We now live in an increasingly toxic world due to the huge amounts of waste we produce and the resultant pollution. This toxicity is affecting our health.

Waste is any material that is not needed by the owner, producer, or processor. Humans, animals, other organisms, and all processes of production and consumption produce waste.

Waste has always been a part of the earth's ecosystem, but its type and scale were such that nature could use the waste in its many cycles. In fact, there is no real waste in nature. The apparent waste from one process becomes the input for another. For example, when leaves fall from trees on the ground, they slowly decompose into soil that nourishes plants and other trees. It is a different story, however, when human beings produce too much waste.

The composition, quantity, and disposal of waste determine the environmental problems it creates. What we dispose of remains in the ecosystem and causes some form of pollution. This pollution can have an impact far away from the point of generation and far removed in time too.

The main categories of wastes are the following:

- Municipal waste, mainly domestic and commercial waste generated in cities.
- Industrial waste.
- Construction waste.
- Agricultural waste.
- Human and animal waste.
- Biomedical waste, mainly from hospitals.
- E-waste or WEEE, which is waste from electrical and electronic equipment.
- Nuclear waste.

Waste could be solid, liquid, or gaseous; it could be hazardous or non-hazardous; it could also be biodegradable or non-biodegradable.

There is no reliable estimate of total waste of all types generated in the world. However, the total amount of municipal waste is estimated to be over 2 billion tonnes a year. The richer the country, the more waste it produces.

Box 6.1 E-waste

Waste from electrical and electronic equipment (WEEE) is commonly known as e-waste. It results from discarded devices like computers, televisions, mobile phones, music systems, and refrigerators. The explosive growth in and the rapid obsolescence of such products have led to a huge increase in e-waste.

E-waste contains many hazardous materials like lead, copper, zinc, aluminium, flame retardants, plastic casings, cables, and others, which can have harmful effects on the environment, if burnt or buried.

Just one mobile phone contains 40 elements of the periodic table, including many metals! Even though the mobile phone is a small device, the waste is large because of the numbers. In 2007, 1.2 billion mobiles were sold in the world and this number is rapidly increasing.

Globally more than 40 million tons of e-waste is generated. In 2007, India's e-waste included 2,75,000 tons from TVs, 1,40,000 tons from PCs, and 2,00,000 tons from refrigerators. China's e-waste is several times more. In addition, both countries receive e-waste as legal and illegal imports for recycling.

NGOs, recycling companies and even manufacturers are now working for the collection and safe recycling of e-waste. The mobile company Nokia, for example, has installed collection boxes in Indian cities. However, the e-waste problem is only getting bigger and bigger.

Table 6.2 shows that, currently, Chinese and Indian citizens generate much less waste per capita than the Americans. What will happen, however, when our consumption levels reach that of the US?

Box 6.2 *Biodegradable and Non-biodegradable Waste*

A biodegradable item breaks down into the raw materials of nature and disappears into the environment. Countless microorganisms together carry out this task.

Some of the products we make are biodegradable, while some are not. Any material that comes from nature will return to nature, as long as it is still in a relatively natural form. Hence, any plant-based, animal-based, or natural mineral-based product can biodegrade. Examples are soap (made from vegetable oil) and plain paper.

Products made from man-made petrochemical compounds generally do not biodegrade. During the production process, we greatly change the original resource. When we throw away the product as waste, the microorganisms are unable to recognize it as coming from a natural resource. In fact, there are perhaps no microorganisms that can break down some of the man-made chemical compounds.

Crude oil will biodegrade in its natural state. Once it is turned into plastic, however, it does not biodegrade. Instead of returning to the cycle of life, such products simply pollute our land, air, and water. Table 6.1 shows how long it takes for some common items of use to biodegrade, when they are thrown into garbage.

Table 6.1 Time taken by waste items to biodegrade

Item	Time to degrade
Cotton rags	1–5 months
Paper	2–5 months
Orange peels	6 months
Woollen socks	1–5 years
Plastic-coated paper cartons	5 years
Leather shoes	25–40 years
Aluminium cans	80–100 years
Plastic bags	450 years
Glass and plastic bottles	Almost forever!

Table 6.2 Municipal waste generation

Country	Municipal waste per person per year (kg)
US	760
China	115
India	70–180 (depending on size of city)

POLLUTION

Pollution is the contamination of the earth's environment with materials that interfere with human health, the quality of life, or the natural functioning of ecosystems. Pollution can occur due to natural causes such as volcanic eruptions, dust storms, and smoke from forest fires. However, our concern here is with man-made pollution, caused by the waste we generate.

Environmental pollution takes place when nature cannot process and neutralize the harmful by-products of human activities within reasonable time without any damage to the ecosystem. Even when nature can handle the pollution, it may take many years for the pollutants to decompose and become harmless. The worst case is radioactive pollution, which may require thousands of years to lose its harmful effect.

Some common categories of pollution are:

- Water pollution (see *Chapter 4*)
- Marine pollution (see *Chapter 8*)
- Soil pollution (see *Chapter 10*)

Box 6.3 What You Should Know about Waste, Pollution, and Environmental Health

- We now live in an increasingly toxic world due to the huge amounts of waste we produce and the resultant pollution. This toxicity is affecting our health.
- Waste has always been a part of the earth's ecosystem, but human activities now generate so much waste that it has now become a major problem.
- There is no 'away' in 'throw away'. The waste we try to dispose off returns to us in a different form.
- Waste pollutes soil, water, and air. This adversely affects human beings, animals, and plants.
- Most of the chemicals used by us are hazardous to us and to the environment.
- Burning fossil fuels in automobiles and industry is the major source of air pollution.
- Noise pollution is on the increase and it has very damaging effects.
- Industrialized countries export their waste, which ends up in countries like India.

- Air pollution (see *Box 6.4*)
- Noise pollution (see *Box 6.5*)

Many items of common use cause pollution. Plastics are perhaps the worst example (see *Box 6.6*)

Box 6.4 Air Pollution

Air in cities gets polluted due to the presence of particulate matter (such as soil, soot, lead, and asbestos), hydrocarbons (like methane and benzene), and compounds of carbon, nitrogen, and sulphur.

Such substances come mainly from:

- Burning of fossil fuels in automobiles, power stations, and industries.
- Construction work.
- Burning wastes of all kinds.
- Natural emissions from animals and decaying organic matter.

At low levels, air pollutants irritate the eyes and cause inflammation of the respiratory tract. Many pollutants also depress the immune system, making the body more prone to infections. Carbon monoxide from automobile emissions can cause headache at lower levels and mental impairment and even death at higher levels.

Particulate matter can reduce visibility, soil clothes, corrode metals, and erode buildings. On a larger scale, air pollution leads to acid rain, ozone layer depletion, and global warming.

Air pollution can be reduced by adopting cleaner technologies, reducing pollution at the source, implementing regulations to make people pollute less, and so on. We can reduce automobile emissions by making cleaner and more fuel-efficient vehicles and improving public transport. Particulate matter in the air can be reduced by cleaning the gases that come out of factories.

Box 6.5 *Noise Pollution*

Noise is defined as unwanted sound and it is an irritant and a source of stress. Most of the noise one hears originates from human activities. The main sources are vehicles, industrial and construction machinery, and special events such as music performances, marriage receptions, religious festivals, and public meetings.

Noise can do physiological or psychological damage if the volume is high or if the exposure is prolonged. Loud, high-pitched noise damages the fine hair cells in the cochlea of the ear. Since the body does not replace damaged hair cells, permanent hearing impairment is caused by prolonged exposure to loud noise.

Noise can also produce other effects like heart palpitation, pupil dilation, or muscle contraction. Migraine headaches, nausea, dizziness, gastric ulcers, and constriction of blood vessels are some of the other possible outcomes. Noise can cause serious damage to wildlife, especially in remote regions, where the normal noise level is low.

We have rules to control noise levels in cities, but we do not implement them. Meanwhile, noise levels in Indian cities are going up.

WASTE, POLLUTION, AND HUMAN HEALTH

Waste and the resulting pollution have been affecting human health in a big way. The effects of some common hazardous chemicals found in waste or used by human beings are described below.

Common Hazardous Chemicals

Most of the chemicals used by humankind are hazardous to some part of, or to the entire, biosphere. Some chemicals affect all species, while others affect only a few. Some cause only minor problems, some can kill instantly. Some are dangerous only when the exposure is long, while others are toxic only in high concentrations. We do not have complete information about the effects of many of the chemicals on humans and other organisms.

Box 6.6 Pollution by Plastic

Developed in the 1860s, plastic has become an indispensable part of our lives. Longevity is its main characteristic, but this very quality has also become a major disadvantage. It is the non-biodegradable nature of plastic that is causing a massive environmental problem.

Plastic bags and bottles now form an increasing proportion of municipal waste and cause many environmental problems. They clog sewage lines and canals and litter public places, gardens, wildlife reserves, and forests. Animals and birds consume plastic items and even die as a result. Disposing of plastic by burning only creates more toxic fumes.

The worst offender is the thin carry bag that easily gets airborne when thrown away. The plastic carry bag has become a common item all over the world. Every year, more than 5 trillion bags are produced, almost all of it being the non-biodegradable variety. About 100 billion bags are thrown away in the US every year.

Certain plastics can be recycled. Many household articles like water containers and buckets are made of recycled plastic. Non-reusable items include carry bags, cups, plates, and magazine wrappers.

There are strong views for and against plastic. The plastics industry cites the advantages of convenience, low cost, employment potential, recycling possibilities, and others. Environmentalists point out the problems listed above and also add in the difficulty of separating the recyclable plastic items in landfills.

The Government of India has banned carrybags made of plastic films of thickness less than 20 micrometers. It is said that manufacturers are therefore busy making bags 21 micrometers thick!

Box 6.7 Find Out and Be Surprised!

1. India manufactures over 4 million tonnes of plastics packaging. Can you guess how much of it ends up as waste soon after usage?
2. In which city will you find Asia's largest market for trade in plastic waste?
3. How much money does New York City spend every day for just carrying away its garbage?

Answers in *Appendix C*

There are broadly two categories of hazardous chemicals that can significantly affect humans and the environment:

- Pesticides and herbicides that are deliberately introduced into the environment for killing pests and unwanted weeds.
- Industrial chemicals that are disposed off as waste or discharged accidentally into the environment: These include organic solvents, waste oil, polychlorinated biphenyls (PCBs), paints, glues, preservatives, and metal residues.

Pesticides

Pesticide is a general term for any chemical that is used to kill unwanted organisms. Modern agriculture and sanitation depend heavily on pesticides. Pesticides have saved millions of lives by killing disease carrying insects. They have also increased our crop yields by eliminating pests that attack plants.

Until the middle of the last century, most pesticides in common use were simple naturally occurring substances like arsenic and nicotine. Then came new chemical compounds like DDT that were much more effective against crop pests and disease carrying insects. The new chemical pesticides, however, remain in the environment much longer than the natural ones. Many are also toxic to humans.

The World Health Organization (WHO) estimates that every year there are three million cases of pesticide poisoning worldwide, resulting in over 2,50,000 deaths. (See Box 6.8 for an account of Endosulfan poisoning in India)

Chemical pesticides also kill non-target species, including the predators of the very organisms they are supposed to eliminate. Many pests are able to develop resistance to the chemicals. This leads to the use of even greater amounts of the chemical or more powerful chemicals, resulting in more problems.

There is a great risk in the manufacture and storage of large amounts of pesticides. Several disasters have already occurred, the Bhopal gas tragedy being the worst.

Many of the new pesticides have been banned in the industrialized countries. However, the same countries continue to manufacture and export them to poor countries, where regulations and enforcement are often lax.

Industrial Chemicals

Some of the common hazardous substances used or formed in industrial processes are described below.

PCBs

PCBs or polychlorinated biphenyls are chemical compounds that are widely used in electrical capacitors, transformers, and others. They are extremely toxic and persist for long in the environment. Humans who mainly consume fish are at risk from PCBs. They can cause liver and nervous disorders. They affect many marine and land animals and this could lead to the extinction of many species.

PCBs have been banned in the industrialized countries. Yet, thousands of tons of PCBs are still present in storage, landfills, or abandoned electrical equipment. A substantial part is in the developing countries with high risk of contamination.

Box 6.8 *The Story of Endosulfan: Poison from the Sky*

Mohana Kumar, a doctor practising in Padre village in Kasaragod District of Kerala, was a puzzled man. Among the patients coming from just two wards of the panchayat, he found very high incidence of cancer, psychiatric problems, mental retardation, epilepsy, congenital anomalies, to name a few.

Kumar began keeping detailed records and felt that the ailments could be due to the pesticide endosulfan. This chemical is effective against a variety of pests that attack crops of cereals, coffee, potato, tea, and vegetables. But it is also highly toxic to humans, birds, and animals. It is easily absorbed by the stomach and lungs and through the skin. Being anywhere near the area of use is enough to get contaminated. Sixty-two countries have banned it, while others have restricted its use.

Since 1976, the government-owned Plantation Corporation of Kerala (PCK) had been carrying out aerial spraying of endosulfan over cashew plantations in Kasaragod including the hills around Padre. The pesticide residues settled on the soil and got washed away into drinking water streams below. Though it is mandatory to cover all water sources like wells, tanks and other water bodies during the spraying of such a toxic pesticide, this was not done.

Kumar and a farmer-journalist Shree Padre found enough published evidence to connect endosulfan with the ailments. Samples of blood, fruits, and animal tissues from Padre were tested at the Indian Institute of Technology, Kanpur (IIT-K) and found to contain extremely high levels of the pesticide. However, PCK and the pesticide manufacturers denied the role of endosulfan in causing the ailments.

Following several studies, court cases, and public protests, the state of Kerala banned the sale and use of endosulfan within its boundary in 2003. The government also started an Endosulfan Relief Cell in 2007 to give some help to the affected people. The state is now seeking a permanent ban on the pesticide by the union government. There are endosulfan victims in Karnataka too and the state government has provided some relief to them.

The production and use of endosulfan continues to poison people and environments across the world. However, a number of countries, led by India, stand in the way of a global ban.

Bio

RUN !
CHOLERA OUTBREAK
DO YOUALSO EAT THESE WITH THAT MASK ON?!
TOXIC

Dioxins

Dioxins are a class of chemical contaminants that are formed during industrial processes including smelting, paper making, and the manufacture of some herbicides and pesticides. They are highly toxic to humans and animals and can cause cancer.

Dioxins are formed during the incomplete incineration of waste and the burning of plastics, coal, or cigarettes. When fuel is partly burnt in a vehicle, dioxins are released. Dioxins get deposited on plants, soil, and water, and thereby enter the food chain.

Lead

Lead is an extremely poisonous metal that accumulates in organisms. Large doses of lead can cause paralysis, blindness, and even death in humans. Lead in the atmosphere contaminates leafy vegetables and fruits.

The lead that we breathe in comes mostly from vehicle exhausts. For long, lead has been added to petrol to prevent the knocking of the engine. Lead-free petrol is now available in India and is mandatory for most new cars. Lead is also found in batteries, paints, bullets, and some alloys. There is a global move to phase out lead from products and processes.

Mercury

Mercury is a liquid metal that is dangerous to humans and animals. While small doses cause headaches, large ones could lead to death. It can also damage the human nervous system. It accumulates easily in the body and is not excreted from the body naturally.

Chemical and plastics industries use large amounts of mercury and release it with the effluents into rivers, lakes, and seas, or dump it on land as waste. This has led to poisoning of fish and people. (See Minamata case below.)

Asbestos

Asbestos is a fibrous silicate mineral. The fibres are woven into a cloth, a binder like cement is added, and the resulting rigid material shaped in many forms. It is a non-corrosive, non-flammable, and non-conducting material, and is inexpensive.

Asbestos is widely used in construction, most commonly as corrugated roof and sometimes also as door or partition. Asbestos wool is used as insulation.

Asbestos is very dangerous to health when the fibres are inhaled. They lodge themselves in the lungs and bronchial tubes. This causes a disease called asbestosis, which affects the respiratory tissues. Chronic shortage of breath and sometimes premature death are the results. The fibres can also cause lung and intestinal cancer.

There is now a worldwide movement against the use of asbestos. There is, however, a large asbestos industry in India. Efforts to ban at least some forms of asbestos in India have not succeeded so far.

Persistent Organic Pollutants

Chemicals such as PCBs, DDT, and endosulfan belong to a group called persistent organic pollutants (POPs). Most POPs are created during human activities, either intentionally or as by-products. They remain in the environment for long, travel far, accumulate in the body, and are not easily detectable. They can cause severe health problems.

Many of the POPs and other hazardous chemicals find their way to our homes.

Industrial Disasters

In many countries, there have been a number of industrial accidents that contaminated the environment for short or long periods. They have caused deaths, severe illnesses, contamination of water, and so on. Here are two examples:

- Minamata Case: In the 1950s and 1960s, some 2,000 people in Minamata, Japan, suffered crippling neurological diseases after eating fish poisoned by mercury wastes discharged into the sea by a local company. More than 400 of them died.
- Love Canal Case: Again in the 1950s, residents of the Love Canal area in New York began developing many health problems. In fact, toxic substances started oozing out in playgrounds and basements. It was discovered that a

Box 6.9 Hazardous Chemicals at Home

There is a cocktail of POPs and other hazardous chemicals in your home. Textiles, plastic goods, detergents, cleaners, insecticides, toiletries, batteries, packaging material, vehicles, daily newspapers, medicines, computers—almost every item that you use has either gone through chemical processes or contains some chemicals. Many of these chemicals also find their way into your food.

Here are some facts for you to think over:

- Around 85,000 chemicals, most of them harmful, are used in everyday cosmetics, even in those claiming to be 'herbal' or 'natural'.
- Mosquito coils and pesticides widely used in homes contain dangerous ingredients.
- There is lead in 90 per cent of paints sold in India.
- Soft toys made of PVC (polyvinyl chloride) contain dangerous additives.
- Cells that you throw away end up in landfills and the deadly chemicals in them leach into groundwater.

chemical company had dumped toxic waste in Love Canal, covered it up, and left. Lois Gibbs, a resident, started a campaign that ultimately led President Carter to declare the neighbourhood as a disaster area. Later, he even set up a superfund to clean up thousands of such dumps in the US.

In addition to industrial disasters, several nuclear accidents and oil spills have taken a heavy toll on human lives and the environment.

Industrial accidents continue to occur in India. Since the Bhopal gas tragedy, over 40 accidents have killed more than 250 people, not counting the deaths in mining accidents. Several of the accidents took place in chemical or gas factories.

Management of Waste

In the industrialized countries, household waste is separated into categories like organic material, paper, glass, other containers, and so on. The separation is often done at home itself, using different bins. In the developing countries, all the waste is collected together, though some cities are trying to persuade the public to separate the waste.

The simplest and most common method used in the cities is to collect and dump the waste in a landfill. These landfills are located just outside the city. There are now thousands of landfills in the world with huge piles of waste. Many countries and cities have run out of space for landfills.

In the poorer countries, ragpickers sift through the waste, collect the reusable and recyclable material, and sell it to the scrap traders. They, in turn, take the material to the recycling units. The ragpickers, the majority of whom are women and children, work under extremely unhygienic conditions. But they provide a great ecological service by manually separating thousands of tons of recyclable waste from the garbage dumps.

Often, the waste in a landfill is burnt away. While this reduces the volume of the garbage, it releases deadly dioxins. Proper incineration of waste needs modern technology and proper management.

How is liquid waste managed? Sewage and industrial effluents are in most cases released directly into water bodies–rivers, lakes, or the ocean. Very often they are not treated before release.

With increasing amounts being generated, management of waste is becoming difficult and expensive. The industrialized countries have found an easier and less expensive method: export the waste to other countries!

Export of Hazardous and Toxic Wastes

Since the economies of the industrialized countries are based on constant growth and consumption, they create huge waste heaps. These countries have strict environmental regulations that make waste management expensive.

The most attractive option for them is to export the waste to developing countries, where disposal is cheap and environmental regulations are lax. The latter need the money and the former want to get rid of the waste. There are many cases of environmental damage caused in the developing countries due to improper management of imported toxic waste.

At any time, there are a number of ships carrying toxic waste prowling on the high seas, ready to dump the waste on an unsuspecting poor country. Alternately, they just dump the cargo somewhere in the middle of the vast ocean.

Huge amounts of wastes are imported into India and the customs authorities are unable to check each consignment for explosive material or toxins. Another major concern is the hazardous waste that is released and generated during ship-breaking.

Ship-breaking

Ship-breaking is a big industry today. Huge vessels, which have served their lives and have been decommissioned, are sent to yards for recycling the parts to the extent possible. India, Taiwan, China, and Bangladesh have large ship-breaking facilities. Alang in Gujarat is one of them.

Most of the ships sent to the yards contain hazardous material such as asbestos, toxic paints, and fuel residues. Ship-breaking has many dimensions: the export of hazardous waste to the developing countries, the environmental problems of handling toxic waste, the safety and health issues of the workers, the social tensions the enterprise creates, the role of the government, and so on.

Hazardous Waste in India

Here are some examples of polluting industries in India:

- More than 2,500 tanneries discharge about 24 mn. cu. m of wastewater containing high levels of dissolved solids and 4,00,000 tons of hazardous solid wastes per year.
- About 300 distilleries discharge 26 million kilolitres of spent wash per year containing many pollutants.
- Thermal power plants discharge 100 million tons of fly ash and this figure is expected to reach 175 million soon. Fly ash contains silicon, aluminium, iron, and calcium oxides and is said to cause silicosis, fibrosis of the lungs, cancer, and bronchitis.

India continues to produce several pesticides banned or restricted in other countries. Examples are DDT, malathion, and endosulfan. One-sixth of the total pesticides used in India are those banned elsewhere. The bulk of the production is from small units that have no technology to treat the toxic and non-biodegradable pollutants generated during the manufacturing process. Almost 50 per cent of the pesticides used in India are for cotton production.

The Ministry of Environment and Forests (MoEF) estimates that 7.2 million tons of hazardous waste is generated annually in India. This will require annually about 1 sq. km of landfill area and Rs 16 billion for treatment and disposal. In addition, industries discharge about 150 million tons of high volume–low hazard wastes every year, which is mostly dumped on open, low-lying land.

Municipal Waste in Indian Cities

The amount of solid waste in Indian cities is growing faster than the population. The total amount of solid waste generated annually in Indian cities was 48 million tons in 1997. It is expected to reach 300 million tons by 2050 and will then require about 170 sq. km of landfill area.

More than 70 per cent of Indian cities do not have adequate waste transportation facilities. Streets piled with garbage, choked drains, and stinking canals are common features of most cities. The garbage that is transported in polluting, old trucks is just dumped in low-lying areas on the outskirts of cities. The choice of the site is more a matter of availability than suitability.

A dangerous practice is the disposal of biomedical waste (from hospitals and clinics) along with municipal waste in dumpsites. The biomedical waste could turn the entire yard infectious. Further, biomedical waste also contains sharp objects like scalpels, needles, broken ampoules, and others, which could injure or infect ragpickers and municipal workers.

In most cities, the solid wastes lie unattended in the dumpsites and attract birds, rodents, insects, and other organisms. As the waste decays, it releases odour and airborne pathogens. Further, the practice of burning the waste creates toxic fumes and spreads the hazardous substances in the air.

Recycling Solid Waste

A good way of dealing with the solid waste problem is recycling, which is the processing of any waste into a new and usable item. There is a large recycling industry in the world. In India, there is a thriving, unorganized recycling industry. The waste and scrap collector buys old newspapers, bottles, used clothes, utensils, scrap, motor oil, and others things and sells them to the recycling units.

Recycling brings multiple benefits:

- By taking away some of the waste, we reduce environmental degradation.
- As against expenditure incurred on disposing of the waste, we make money out of the waste material.
- We save energy that would have gone into waste handling and the making of products.

Safe and profitable technologies for recycling paper, glass, metals, and some forms of plastic are available. Biogas can be produced from landfill waste and paper factories can certainly recycle their waste.

Recycling is not a solution for all waste material. In many cases, the technologies are not available or unsafe. In other cases, the cost of recycling is too high.

You can take many steps to reduce waste and pollution. Your health will improve and you will save money too!

It is far better to prevent generation of waste than to produce waste and then try to 'manage' it. We cannot simply throw away waste: *'There is no away in throw away.'*

Think about it: About 400 tons of waste was produced in India during the 30 minutes it took you to read this chapter!

Box 6.10 *What You Can Do to Minimize Waste and to Manage Waste Better*

- Reduce your consumption. Buy only what is necessary. Do not be taken in by telemarketing, advertisements, and 'sale' campaigns.
- Examine every item of waste that is generated in your home. Find out where it came from and where it is headed. In each case, try to prevent the generation of that waste item. If it cannot be prevented, can it be reused or recycled in any way? Is there an alternative to throwing it in the garbage bin?
- Segregate waste: Keep separate bins for degradable and non-degradable waste. Compost kitchen waste and get organic manure for your garden.
- Reuse every possible item at home—paper, plastic bags, cards, envelopes, wood, and so on. You can make artistic items out of waste.
- Minimize the use of paper:
  - Use both sides of each sheet.
  - Collect and reuse sheets printed on one side.
  - Minimize computer printouts.
  - Reduce junk mail by writing to firms to take your address off their mailing lists.
  - Send greetings by email and not through fancy cards.
  - Do away with paper tissues; go back to handkerchiefs and cloth napkins.
- Sell old paper, metal, and other things, to the waste trader.
- Buy recycled paper products.
- Avoid heavily packaged products.
- If you are buying a car, buy a small and efficient one.
- Do not buy soft drinks in metal containers
- Minimize packaging waste by taking your own cotton bags to the market and the grocery store. Do not buy pre-packaged items. Buy reusable or refillable containers.
- Avoid supermarkets and if you do shop in them, take your own bag. Refuse the large plastic carry bag offered by the shop. Even better, support the small store, which uses old newspaper to pack groceries. Try to persuade supermarkets in your area to offer a small discount to shoppers who bring their own bags; tell them that they can benefit by spending less on bags and advertising their 'green' concerns.
- Buy durable products that will last long.

Box 6.11 What You Can Do to Minimize the Use Of Chemicals and Hazardous Substances

- Educate yourself about hazardous and dangerous chemicals. Examine every item you use at home and try to find out if it contains such chemicals or if it has been processed using such chemicals. If the answer is yes, try to replace the item with a more eco-friendly and safer product.
- Avoid the use of chemical mosquito repellents in your home and in those of your friends. Covering windows with nets and using a mosquito net are far safer methods of protection against mosquitoes. If you must use a repellent, choose natural substances like lemongrass (citronella) oil.
- Do not buy items like naphthalene balls, insect sprays, floor cleaners, detergents, drain cleaners, paint thinners, etc. Try to use safer natural alternatives like ammonia, bleaching powder, baking soda, and vinegar.
- Use natural colours for Holi. For Ganesh puja, buy only mud idols without any chemical paints. After the puja, immerse the idol just in a bucket of water and not in a pond or lake.
- Buy rechargeable batteries.
- Do not throw away toxic material like batteries, thermometers, and insecticides as garbage. Find out if the manufacturers will take them back or if anyone will recycle them.
- Do not throw unwanted medicines or motor oil down the drain. If the medicines have not expired, donate them to any charitable hospital or voluntary organization that accepts them. There is a thriving industry in cities that recycles motor oil.
- Do not use chemical fertilizers and pesticides for your garden.
- Buy natural products like earthen cups, leaf plates, bamboo dustbins, etc.

Box 6.12 What You Can Do to Minimize Air and Noise Pollution

- Ensure that your vehicle does not spew out excessive smoke.
- Discourage your friends or neighbours from lighting bonfires. Never burn tyres or plastics, which release deadly dioxins and poisonous gases.
- Say 'no' to crackers during Diwali and other occasions.
- When a group creates too much noise in your neighbourhood, especially at night, try to persuade them to stop or reduce the volume. If they do not respond, feel free to go the police.

Box 6.13 A Poem by Dr Seuss*: The Lorax

The children's book *The Lorax* by Dr Seuss is a great environmental story and you may like it for its message and its wacky verse. Here is an extract from the book:

Then again he came back! I was fixing some pipes
when that old nuisance Lorax came back with more gripes.
I am the Lorax, he coughed and he whiffed.
He sneezed and he snuffled. He snarggled. He sniffed.
Once-ler! he cried with a cruffulous croak.
Once-ler! You're making such smogulous smoke!
My poor Swomee-Swans ... why, they can't sing a note!
No one can sing who has smog in his throat.

And so, said the Lorax,
—please pardon my cough–
they cannot live here.
So I'm sending them off.
Where will they go?...
I don't hopefully know.
They may have to fly for a month ... or a year ...
To escape from the smog you've smogged-up around here.

* (1904–1991); Theodore Seuss Geisel, American poet, writer, and cartoonist.

END PIECE

Robert Orben (b. 1927), American magician and professional comedy writer, once said:

There's so much pollution in the air now that if it weren't for our lungs there'd be no place to put it all.

EXPLORE FURTHER

Books and Articles

Dengel, Lucas, 2001, *The Gita of Waste*, Auroville, Tamil Nadu: Auroville Health Centre.

Epstein, Samuel S. and Randall Fitzgerald, 2009, *Toxic Beauty: How Cosmetics and Personal-Care Products Endanger Your Health and What You Can Do About It*, Dallas, Texas: Ben Bella Books.
Gibbs, Lois Marie, 1982, *Love Canal: My Story*, New York: State University of New York Press.
——, 1995, *Dying from Dioxin*, Boston: South End Press.
Jamwal, Nidhi, 2003, 'e-Waste', *Down to Earth*, Vol. 12, No. 12 (15 November), pp. 50–1.
Joshi, Sopan, 2001, 'Children of Endosulfan', *Down to Earth*, Vol. 9, No. 19 (28 February), pp. 28–35.
Lapierre, Dominique and Javier Moro, 2001, *It Was Five Past Midnight in Bhopal*, New Delhi: Full Circle Publishing.
Mukherjee, Swaroopa, 2002, *Bhopal Gas Tragedy: The Worst Industrial Disaster in Human History*, Chennai: Tulika Publishers.
Smith, Rick and Bruce Lourie, 2009, *Slow Death by Rubber Duck: How the Toxic Chemistry of Everyday Life Affects Our Health*, Toronto: Alfred A. Knopf.
Toxics Link, 2010, *Our Toxic World: A Guide to Hazardous Substances in our Everyday Lives*, New Delhi: Sage Publications and Toxics Link.
Wolf, Gita, Anushka Ravishankar, and Orijit Sen, 1999, *Trash: On Ragpicker Children and Recycling*, Chennai: Tara Publishing and Bangalore: Books For Change.
Yadav, Kushal P.S. and S.S. Jeevan, 2002, 'Endosulfan Conspiracy', *Down to Earth*, Vol. 11, No. 4 (15 July), pp. 25–34.
For articles on various aspects of the Bhopal disaster and its aftermath, read the articles in *Seminar*, No. 544 (December 2004), an issue devoted to Bhopal under the title *Elusive Justice*.

Websites

Bhopal disaster: www.bhopal.net
www.bhopal.org
www.studentsforbhopal.org
Center for Health, Environment, and Justice (CHEJ): www.chej.org
Minamata: http://corrosion-doctors.org/Elements-Toxic/Minamata-1.htm
www1.umn.edu/ships/ethics/minamata.htm

Films

Bhopal Express (Mahesh Mathai, 1999)
For more films, see http://film.bhopal.net

7 A WORLD WITHOUT LIFE
Forests and Biological Diversity

The best time to plant a tree was twenty years ago.
The second best time is now.

—ANONYMOUS

THE STORY OF THE GREAT INDIAN BUSTARD AND OTHER BIRDS: WINGED WONDERS, DISAPPEARING SPECIES

The Great Indian Bustard stands majestic with a height of one metre. It is a symbol of the Indian grasslands and is the state bird of Rajasthan. Today, however, it is on its way to extinction. It is one of the 78 globally threatened bird species.

The bustard is found in the Thar Desert and five other states. The short grass plains where the bustard lives are disappearing everywhere. They are being replanted and overgrazed. When the forest departments plant trees on the grasslands, they destroy the bustard's habitat.

The Bombay Natural History Society (BNHS) and the Indian Bird Conservation Network (IBCN) have started conservation programmes to save the bustard. In 2000, they organized a 350 km *padayatra* from Bikaner to Jaisalmer, explaining to the local people the importance and plight of the bustard.

Bird species are under threat all over the world. More than 1,200 bird species, out of 9,800 known ones, are likely to become extinct within this century. Over 900 species are already either endangered or critically endangered. About 128 species have vanished over the last 500 years, and 103 of them since 1800.

Birds perform valuable ecosystem services like seed dispersal, insect and rodent control, scavenging, and pollination. Many birds are sentinel species that warn us of impending or current problems. They could indicate high acidity levels in water, chemical contamination, arrival of new diseases, and effects of global warming.

Colombia, which has more than 1,800 species, is the most bird-diverse country. India, with about 1,250 species, ranks among the top ten bird-diverse countries of the world.

The major threats to birds include habitat loss, hunting and capture, oil spills, pesticides, and herbicides. Bird lovers and bird watchers all over the world are trying to save the winged wonders. The future, however, looks bleak for birds everywhere.

WHAT DOES THE STORY OF DISAPPEARING BIRDS TELL US?

We now live in a world that is rapidly losing its forests and many species of plants and animals.

Every species plays a role in the earth's ecosystem. When a bird species goes extinct, many food chains are cut and we do not even know the consequences. Not only do we have to save birds, we should also attend to the problems that cause birds and other species to disappear.

NUMBER OF SPECIES IN THE WORLD

We do not know exactly how many species inhabit this earth. Estimates range from 4 million to 100 million. The best guess is 10–14 million. Most of the species in the world are insects and microorganisms not visible to the naked eye. Every single species, however, has a role to play in the biosphere.

So far about 1.9 million species (not including bacteria) have been identified, named, and catalogued. These include 2,70,000 plant species, 45,000 vertebrates, and 9,50,000 insects. Even among these, we understand the exact roles and interactions of just a small number of species. Roughly 10,000 new species are identified every year.

GEOGRAPHICAL SPREAD OF BIODIVERSITY

The vast majority of all species are in the developing countries. About 50 to 75 per cent of all species are to be found in the tropical forests that account for just 6 per cent of the land area. A handful of soil in a tropical forest contains hundreds of species and more than a million individual organisms.

Box 7.1 What You Should Know about Forests and Biodiversity

- We now live in a world that is rapidly losing its forests and many species.
- There are millions of species on earth, but we have studied only a small proportion of them.
- The earth's biodiversity is its primary life support system and vital to our survival. Forests, for example, provide us with invaluable products and services.
- Most of the biodiversity exists in the tropical developing countries and India is one of the mega-diversity countries of the world.
- Five mass extinctions have occurred so far on earth due to natural processes. But the sixth one, currently under way, is due to human-induced reasons.
- In India and the rest of the world, hundreds of known species are threatened with extinction and many are endangered. This affects the complex web of life with unknown consequences.
- Forest cover on earth is depleting rapidly, especially in the tropics and this will have serious consequences for the environment.
- There are a range of international conventions and agreements to promote conservation of forests and biodiversity, but implementation has been poor.
- Measures to conserve biodiversity include the establishment of nature reserves, seed banks, zoos, and botanical gardens, as well as the involvement of local communities.
- We are now realizing the importance of traditional and local knowledge of forests and biodiversity and the need for recording it.

In the tropics and the sub-tropics there was always evolutionary activity giving rise to a rich biodiversity. Biodiversity is less in the colder northern regions because the recurrent ice ages there slowed down the proliferation of life forms.

Almost all the plants eaten today in Europe originated in the developing countries. The genetic diversity needed to maintain the world agricultural system is mainly in these places. Most of the medicinal plants too are also found only in the developing countries.

The density of species is very high in the southern hemisphere. About 70 per cent of the world's species is found in just 12 countries: Australia, Brazil, China, Colombia, Costa Rica, Ecuador, India, Indonesia, Madagascar, Mexico, Peru, and the Democratic Republic of Congo. The entire Hindu Kush-Himalayan belt has as many as 25,000 plant species, comprising 10 per cent of the world's flora.

EXTINCTION OF SPECIES

By extinction, we mean the complete disappearance of a species, that is, not a single member of the extinct species is found on earth. It is an irreversible loss and is called biological extinction.

There are mainly two ways in which species become biologically extinct. Background extinction refers to the gradual disappearance of species due to changes in local environmental conditions. The rate of background extinction has been generally uniform over long geological periods. On occasions, however, a second type, called a mass extinction, has occurred on earth.

Mass Extinction

Mass extinction is characterized by a rate of disappearance significantly higher than the background extinction. It is often a global, catastrophic event, with more than 65 per cent of all species becoming extinct over some millions of years. This is a brief period in geological terms, compared to 4.6 billion years the earth has existed.

Scientists believe that there have been five mass extinctions over the past 500 million years. In each case, the character of ecological communities changed dramatically. After each such extinction, it took 20 to 100 million years for the global biodiversity to recover.

The most severe extinction occurred about 225 millions years ago, when 95 per cent of marine species vanished. The most famous (and the last) mass extinction, however, took place about 65 million years ago, when the giant dinosaurs, which had ruled the world for 140 million years, were wiped out.

Both biological and environmental factors could have led to mass extinctions. The suggested causes include global cooling, falling sea levels, predation, competition, and a hit by a giant comet or asteroid.

Box 7.2 Find Out and Be Surprised!

How long has the cockroach as a species existed on earth?

Answer in *Appendix C*

Loss of the World's Species

Before humankind became very active, the world was losing annually one out every million species. In early twentieth century, we were perhaps losing one species a year. Now, we are losing one to hundred species a day and soon the rate of extinction is likely to be one thousand each day. These are estimates, but it is definite that we are losing species rapidly. Box 7.2 gives the main causes of biodiversity loss.

Over the next 50 to 100 years, as population and resource use grow exponentially, the rate of biodiversity loss will also increase sharply. More than 20 per cent of the current species could be gone by 2030 and 50 per cent by the end of the century.

Biologists are convinced that we are in the midst of the sixth mass extinction, caused by human activities. They are concerned at the disappearance of sensitive species like birds, frogs, and bees.

Box 7.3 Causes of Biodiversity Loss

The major causes for the decline in biodiversity are the following:

- Habitat loss and degradation: Destruction of tropical forests, coral reefs, wetlands, and so on; Pollution of freshwater streams, lakes, and marine habitats.
- Habitat fragmentation: Due to human impact, many large, continuous areas of habitat are being reduced in extent or divided into patchwork of isolated fragments.
- Commercial hunting and poaching.
- Introduction of non-native species: When a non-native species is introduced in an ecosystem that has no predators, competitors, parasites, or pathogens to control its numbers, it can reduce or wipe out many local species.

IMPACT OF BIODIVERSITY LOSS

The poor people in the developing countries, who are dependent on biodiversity for their daily survival, will feel the impact first. Soon, however, the industrialized countries will also start experiencing the effects. Most of their food crops, medicines, textiles, spices, dyes, and paper originate from plants in the developing countries.

Many scientists and environmentalists believe that the mass extinction that is now underway could be the biggest of our environmental problems. Almost all the other problems are potentially reversible. The loss of biodiversity is not.

BIODIVERSITY CONSERVATION

In biodiversity conservation, we study how human activities affect the diversity of plants and animals and develop ways of protecting that diversity. Conservation ranges from protecting the populations of a specific species to preserving entire ecosystems.

There are two main types of conservation. In-situ (on-site) conservation tries to protect species where they are, that is, in their natural habitat. It primarily means the establishment of protected natural parks and reserves in natural areas that have high biodiversity. Ex-situ (off-site) conservation attempts to preserve and protect the species in a place away from their natural habitat. It includes the storage of seeds in banks, breeding of captive animal species in zoos, and setting up botanical gardens, aquariums, and research institutes. In general, in-situ conservation is more cost-effective. In many cases, however, the ex-situ approach may be the only feasible one.

We are now realizing the importance of traditional and local knowledge of forests and biodiversity and the need for recording it. We also realize the need for involving local communities in conserving biodiversity and forests.

CONVENTION ON BIOLOGICAL DIVERSITY

The UN Convention on Biological Diversity (CBD) came into force in 1993. As of February 2010, 193 countries are parties to the convention.

The convention has three main goals:

- The conservation of biodiversity.
- Sustainable use of the components of biodiversity.
- Sharing the benefits arising from the commercial and other utilization of genetic resources in a fair and equitable way.

Under the convention, governments undertake to conserve and sustainably use biodiversity. They are required to develop national biodiversity strategies and action plans, and to integrate these into broader national plans for environment and development.

MEDICINE
PILL
W's

WHAT'S SO 'GREAT' ABOUT BEING A GREAT INDIAN BUSTARD?

BIODIVERSITY IN INDIA

India is one of the 19 mega-biodiversity countries and the rich biodiversity is attributed to the presence of a variety of ecosystems and climates. So far, about 70 per cent of the total area has been surveyed for biodiversity assessment. More than 45,000 wild species of plants and 81,000 wild species of animals have been identified. Together they represent 6.5 per cent of the world's biodiversity. Many biologically-rich areas like the north-east have not been fully explored.

Almost 18 per cent of all plants found in India are unique to the country. At least 166 crop species and 320 species of wild relatives of crops originated in India. We also have a large variety of domesticated species. For example, the number of rice varieties alone ranges from 50,000 to 60,000.

Threats to the Biodiversity of India

At least 10 per cent of India's plant species and a larger percentage of its animal species are threatened. The cheetah and the pink-headed duck are amongst the well-known species that have become extinct. More than 150 medicinal plants have disappeared in recent decades. About 10 per cent of flowering plants, 20 per cent of mammals, and 5 per cent of birds are threatened.

Hundreds of crop varieties have disappeared and even their genes have not been preserved. The Green Revolution encouraged farmers to plant the new high-yielding varieties (HYVs) replacing the indigenous ones. In place of thousands of varieties of rice, Indian farmers now plant just a small number of HYVs.

The biodiversity hotspots in India are the eastern Himalayas and the Western Ghats. Both contain a large number of unique species, which are now under threat.

Conservation of India's Biodiversity

The first protected area in the country was the Corbett National Park, established in 1936. Currently, there are more than 500 national parks, sanctuaries, and biosphere reserves covering about 5 per cent of the land. There are 35 botanical gardens and 275 zoos, deer parks, safari parks, and aquaria.

In May 1994, India became a party to CBD. In January 2000, the government released the National Policy and Action Strategy on Biodiversity. This document seeks to consolidate the on-going efforts of conservation and sustainable use of biodiversity and to establish a policy and programme regime for the purpose.

The Indian Parliament passed the Biodiversity Bill in December 2002. The main intent of this legislation is to protect India's rich biodiversity and associated knowledge. It seeks also to prevent the use of our biodiversity by foreign individuals and organizations without sharing the benefits with us. One thrust area is the conservation of medicinal plants.

FORESTS

Let us now turn to forests, which are the main storehouses of biodiversity. Forests provide us with a range of products and services.

Box 7.4 Value of Forests

Forests provide us with many goods and ecosystem services:

- Environmental stability: Forests keep the natural environment stable through ecosystem services such as the following:
  - Regulation of global carbon and nitrogen cycles: For example, they absorb carbon and release oxygen
  - Regulation of local climate: Trees absorb water from the soil and release it from the leaves. This provides moisture for the clouds, resulting in rainfall.
  - Conservation of soil: Forests hold soil in place, preventing erosion and mudslides.
  - Conservation of water: Forests absorb, hold, and slowly release water. This ensures flow even during dry periods and controls floods and droughts.
  - Conservation of biodiversity: Forests are the habitats for thousands of species of plants and animals.
- Goods and economic resources: Forests provide us with a range of useful goods:
  - Timber and bamboo for construction, paper production, and fuelwood.
  - Fruits, nuts, honey, essential oils, resins, and others.
  - Medicinal plants and spices.
  - Grazing for livestock.
  - Employment opportunities for people.
- Aesthetic values: The beauty and natural atmosphere of forests are a source of recreation and spiritual solace for the people who visit them.

State of the World's Forests

It is difficult to assess the extent of the world's forests. Data supplied by countries are often not reliable, though satellite imagery is being increasingly used to verify data from ground surveys. The definition of a forest also varies from one assessment to another.

The key findings of the UN FAO (Food and Agriculture Organization) Global Forest Resources Assessment 2010 were:

- Forests cover 31 per cent of total land area. The world's total forest area is just over 4 billion ha., which corresponds to an average of 0.6 ha. per capita.
- The rate of deforestation shows signs of decreasing, but is still alarmingly high. Deforestation (mainly the conversion of tropical forest to agricultural land) shows signs of decreasing in several countries but continues at a high rate in others. Insect pests and diseases, natural, disasters and invasive species are causing severe damage in some countries.
- South America and Africa continue to have the largest net loss of forest. In Asia, the forest area increased between 1990 and 2010 as a result of large-scale afforestation efforts, particularly in China.
- Primary forests account for 36 per cent. The area of planted forest accounts for 7 per cent of total forest area. Large-scale planting of trees is significantly reducing the net loss of forest area globally.
- Legally established protected areas cover an estimated 13 per cent of the world's forests.
- Forests store a vast amount of carbon.
- Almost 12 per cent of the world's forests are designated for the conservation of biological diversity, and 8 per cent of the world's forests have protection of soil and water resources as their primary objective.
- Almost 30 per cent of the world's forests are primarily used for production of wood and non-wood forest products. Wood removals increased between 2000 and 2005 and fuel wood accounted for about half of the removed wood.
- Around 10 million people are employed in forest management and conservation, but many more are directly dependent on forests for their livelihoods.
- Significant progress has been made in developing forest policies, laws and national forest programmes. More than 80 per cent of the world's forests are publicly owned, but ownership and management of forests by communities, individuals, and private companies is on the rise
- The number of university students graduating in forestry is increasing.
- There are many good signs and positive trends towards sustainable forest management at the global level.

Impact of Human Activities and Natural Forces on the Forests of the World

The impact of human activities as well as the forces of nature on the forest ecosystem includes the following:

- The clearing and burning of the forests for agriculture, cattle rearing, and lumber result in loss of biodiversity, extinction of species, soil erosion (resulting in the loss of vital topsoil), and disturbance of the carbon cycle (leading to global warming).
- The clear cutting and conversion of forest land on hilly areas for agriculture, plantations, and housing leads to landslides and floods that affect people in the forests and on the plains. It also increases siltation of rivers.
- Many forests have been affected by acid deposition originating from industries.
- The harvesting of old growth forests destroys crucial habitat for endangered species.
- Pesticide spraying to control insects in forest plantations leads to poisoning all the way up the food chain and to unintended loss of predatory hawks, owls, and eagles. This in turn leads to the increase of the pest population.
- Dams built in forest areas for hydropower and water drown huge areas, destroying species.
- In wilderness areas like the Arctic, oil exploration and military activities disrupt the ecosystem, contaminate areas, and lead to the decline of species.

Saving the World's Forests

The 1999 Report of the World Commission on Forests and Sustainable Development (WCFSD) gave the following recommendations for conserving forests:

- Stop the destruction of the earth's forests: their material products and ecological services are severely threatened.
- Use the world's rich forest resources to improve life for poor people and for the benefit of forest-dependent communities.
- Put the public interest first and involve people in decisions about forest use.
- Get the price of forests right, to reflect their full ecological and social values, and to stop harmful subsidies to lumber companies.
- Apply sustainable forest management approaches so we may use forests without abusing them.
- Plan for the use and protection of whole landscapes, not the forest in isolation.
- Make better use of knowledge about forests, and greatly expand this information base.
- Accelerate research and training so that sustainable forest management can quickly become a reality.

Box 7.5 *What You Can Do to Save Forests and Conserve Biodiversity*

- Buy your food from local markets and small traders and not from supermarkets. The big supermarkets (and fast food chains) tend to offer a small variety of species and thus endanger biodiversity. They also drive out small local businesses. You may be attracted by their current low prices, but ultimately it will all boomerang on you.
- Buy local varieties of food items, fruits, and vegetables in preference to those that have come from abroad or from a place very far from your town. Buying local species ensures that the diversity is conserved.
- Save the forests by saving paper.
- Contribute to organizations that promote tree planting.
- Plant trees—in your compound, neighbourhood, park, streets, on the denuded slopes of a hill, and elsewhere.
- Change your food habits and start eating indigenous varieties of rice and wheat. You will be conserving biodiversity and at the same time eating more healthy food.
- Do not buy any product made from killing endangered animal species. This includes products made from ivory.
- Eat less meat, or even better, become a vegetarian or a vegan.
- If you have some land or even just a terrace, grow vegetables using organic methods. Or, start a community garden for vegetables and herbs.
- Buy only organically grown food; if it is not available locally, form a consumer group to procure and distribute organic food.

- Take bold political decisions and develop new civil society institutions to improve governance and accountability regarding forest use.

There is no sign of these recommendations being implemented. If we do not act soon, the whole world may go the way of the Easter Island (see Chapter 2). But you can act as an individual to save forests and biodiversity.

Just to Let You Know: During the 30 minutes it took for you to read this chapter, about 1,000 hectares of rainforest have been destroyed!

Box 7.6 Hopeful Signs

There are remarkable stories from all over the world of individuals and groups going to great lengths to save or plant trees and to conserve biodiversity. Some examples are:

- Chipko: The people's movement to save the forests of Tehri Garhwal. It began in 1974 in the village of Reni, where the women resisted the felling of trees by a contractor. (See Chapter 13 for the full story.)
- The Green Belt Movement in Kenya: Wangari Maathai encouraged thousands of women to plant trees. In return, they received much-needed fuel wood. For her work, Wangari was awarded the Nobel Peace Prize in 2004.
- The planting of roadside avenue trees in Karnataka by Thimmakka and Chikkanna, a poor labourer couple.
- The unbelievable two-year tree-sit by Julia Butterfly Hill to save the redwood trees of California in 1997–9.
- The Foundation for the Revitalisation of Local Health Traditions (FRLHT), Bangalore, which aims to revitalize the Indian medical heritage by conserving natural resources used by Indian systems of medicine and promoting the transmission of traditional knowledge of health care.

Box 7.7 A Poem by Lao Tzu

Here are the words of Lao Tzu, the Chinese philosopher of the 6th century BC, from his book *Tao Te Ching*:

In harmony with Tao
The sky is clear and spacious
The earth is solid and full
All creatures flourish together
Content with the way they are
Endlessly repeating themselves
Endlessly renewed

When humanity interferes with Tao
The sky becomes filthy
The earth becomes depleted
The equilibrium crumbles
Creatures become extinct.

END PIECE

Reflect on what the French writer, politician and diplomat, Chateaubriand (1768–1848) said:

Forests precede civilizations, deserts follow them.

EXPLORE FURTHER

Books and Articles

Hill, Julia Butterfly, 2000, The Legacy of Luna, San Francisco: HarperCollins.

Leakey, Richard and Roger Lewin, 1996, *The Sixth Extinction: Patterns of Life and the Future of Humankind*, New York: Anchor Books, Random House.

Maathai, Wangari, 2007, *Unbowed: A Memoir*, New York: Anchor Books.

Tyler, Patrick E., 2004, 'For an Environment of Peace', *Frontline*, Vol. 21, No. 22 (5 November), pp. 46–8.

Wilson, Edward O., 2003, *The Future of Life*, New York: Vintage Books.

Youth, Howard, 2003, *Winged Messengers: The Decline of Birds*, World Watch Paper 165, Washington, D.C: World Watch Institute.

Websites

Audobon Society (devoted to birds): www.audobon.org

Biodiversity conservation: www.biodiversity911.org

Foundation for the Revitalisation of Local Health Traditions: www.frlht-india.org

Greenbelt Movement: www.greenbeltmovement.org

Thimmakka: http://www.outlookindia.com/article.aspx?207401
www.goodnewsindia.com/Pages/content/inspirational/thimmakka.html

UN FAO Global Forest Resources Assessment 2010: www.fao.org/forestry/fra/fra2010/en/

Wangari Maathai: www.nobelprize.org/peace/lauraetes/2004/index.html

World Commission on Forests and Sustainable Development (WCFSD): www.iisd.org/wcfsd/

8 A DYING OCEAN
Polluted Seas and Coasts

Give a man a fish, and he can eat for a day.
But teach a man how to fish,
and he'll be dead of mercury poisoning inside of three years.

—ANONYMOUS

THE STORY OF THE GULF OF MANNAR: UNIQUE ECOSYSTEM IN PERIL

The island of Kurusadai, near Rameswaram, is a biologist's paradise. It contains dense foliage, shallow coral reefs, and aquatic life with an extraordinary diversity of form and colour. This is just one example of the marine life in the Gulf of Mannar, one of the world's richest marine biodiversity regions.

The Gulf of Mannar Biosphere Reserve (GOMBR) is located in the south-eastern tip of Tamil Nadu extending from Rameswaram to Kanyakumari. It covers an area of 10,500 sq. km, includes 21 islands, and contains over 3,600 species of fauna and flora. They include mangroves, seagrass beds, algae, coral reefs, many fish species, marine turtles, dolphins, dugongs, and many migratory birds.

This biological paradise is now in trouble. The habitat is getting degraded, the marine species are threatened, and people's livelihood seriously affected. The three main threats to the ecosystem are the destruction of the marine resources by bottom trawling, the mining of corals, and the pollution from industries.

The trawlers scrape the seabed using huge nets weighed down by chains. In the process, they destroy the breeding grounds of a large number of marine species. Since the nets are of fine mesh, they also bring up large amounts of smaller fish (bycatch), thus decimating marine life. The traditional fishers, on the other hand, use small boats and baiting methods that do not harm non-target species.

There is a perpetual conflict between the two groups. The trawlers are supposed to operate beyond three nautical miles from the shoreline and the fishers within this distance. Each group accuses the other of violating the distance restriction. Mutual arrangements regarding times and days of operation do not work.

Illegal coral mining goes on unchecked in the area. Several hundred boats mine huge quantities of coral every day using levers and dynamite to break up the reefs. The catch is sent to cement companies as raw material. About 65 per cent of the coral reefs in the area are already dead, mostly due to human interference. The dugongs have disappeared and many marine species are threatened.

Over 1,00,000 people in 90 villages and hamlets in the area depend on fishing for their livelihood. Over the years, the variety and catch of fish has steadily gone down. As a result, there is great indebtedness among the fishermen.

The Gulf of Mannar was declared a biosphere reserve in 1989 and was brought under the Man and the Biosphere Programme of UNESCO (United Nations Educational, Scientific and Cultural Organization) in 2001. This implies that fishing or any other activity is prohibited in the area. Such a restriction is unfair and unenforceable, as it goes against the traditional rights of the local fishers. The only way to save the biosphere is by involving the local people as partners.

The Global Environmental Facility (GEF) has now provided support to the biosphere reserve through the GOMBR Trust. This body is responsible for the coordination of a management plan for the biosphere reserve in cooperation with government agencies, private entrepreneurs, and local people's representatives. Priority is being given to community-based management.

The director of GOMBR Trust said in 2010 that the formation of the trust and constant interaction with the locals had prevented mining of coral reef but the coastal communities were still smuggling sea horses, sea cucumber, and seaweeds. The biosphere reserve is still in a critical condition.

WHAT DOES THE STORY OF THE GULF OF MANNAR TELL US?

We now live in a world in which the great ocean, inland seas, and the coasts are being exploited and polluted beyond measure.

The GOMBR story is again a case of a unique natural environment, population pressure, people's livelihood issues, and conservation efforts. The story tells us how valuable the ocean is and how human impact is adversely affecting the ocean ecosystems.

Box 8.1 What You Should Know about the Ocean and the Coastal Zone

- We now live in a world in which the great ocean, the inland seas, and the coasts are being exploited and polluted beyond measure.
- The ocean is rich in biodiversity and provides us with many economic resources and ecosystem services.
- Human activities pose great threats to the marine ecosystem. They are polluting the ocean and threatening the marine living resources.
- Coral reefs and mangroves play vital roles in the marine ecosystem, but they are in serious decline all over the world.
- Many large inland seas and freshwater lakes are facing severe environmental problems.

- The coastal zone with its immense biodiversity is being degraded by development and population pressure.
- A number of international agreements and national regulations aim at conserving the resources of the coastal zone and the open ocean.

THE OCEAN

This planet should really be called the ocean and not the earth, because the ocean covers 71 per cent of the planet's surface. The ocean has an area of about 361 million sq. km, an average depth of 3.7 km, and a total volume of 13,47,000 mn. cu. km.

Although we talk of different oceans in the world, there is just a single global ocean. For convenience, we have divided the ocean into the Pacific, the Atlantic, the Indian, and the Arctic. The largest geographical division is the Pacific Ocean that covers one-third of the earth's surface and contains more than 50 per cent of its water. The Pacific also has the deepest part of the ocean, the Mariana Trench, with a maximum depth of 11,000 m, which is 2,200 m more than the height of Mt Everest!

Box 8.2 Resources and Services from the Ocean and the Coastal Zone

Economic resources and services:

- Food items: fish, other marine animals, sea weed, and other items.
- Energy: oil and natural gas, ocean thermal energy, wave energy.
- Minerals: manganese, gold, silver, zinc, copper, and others.
- Medicines.
- Building materials: lime from shells, cement from corals.
- Transportation.
- Recreation.
- Employment.

Ecosystem services:

- Regulation of climate and rainfall.
- Cycling of nutrients.
- Absorption of carbon dioxide.
- Waste treatment and dilution.
- Habitat and nursery for many species.
- Storage of biodiversity and genetic resources.
- Protection from storms: Mangroves, coral reefs, coastal wetlands, and others, act as barriers and reduce the impact of storms.

The global ocean is important to us in many ways. The biodiversity and resource-base of the ocean far exceeds that of the land. The ocean contains more than 2,50,000 species of plants and animals. This could well be an underestimation, since we have explored just a small part of the ocean. The ocean provides us with many ecosystem services and economic resources.

INLAND SEAS

Inland seas are large saline lakes that occur on all continents. They include the Caspian Sea, Dead Sea, Black Sea, and Aral Sea. They are sensitive ecosystems and their environmental importance is only now becoming clear. Most of the large inland seas of the world are facing environmental degradation. They were once healthy waters used for fishing, transport, and other purposes. Now, however, human activities like overfishing, pollution, and diversion of inflow are decimating many of these water bodies.

The Aral Sea, on the border between Uzbekistan and Kazakhstan, was once the world's fourth largest freshwater lake. It was spread over 68,000 sq. km in 1960. Over the past 50 years, however, it has lost over 75 per cent of its volume and over 50 per cent of its area, due to human activities. Efforts are now on to save this inland sea.

IMPACT OF HUMAN ACTIVITIES ON THE OCEAN

The apparent vastness of the ocean gives us the impression that it can accommodate all kinds of uses and can absorb any amount of waste. We now know, however, that the ocean is a finite resource and that human activities are already having an extremely adverse impact on it.

The deep seabed has mineral as well as biological resources and many countries are now developing techniques to tap this wealth. The insatiable demand for petroleum has led to large-scale extraction of oil from the ocean with many negative consequences. The main shipping lanes of the ocean now carry very heavy traffic with consequent problems like large oil spills from giant tankers.

The dumping of hazardous waste, including nuclear material, into the ocean is raising pollution levels to great heights. Near the shore, heavy discharges from industry and sewage systems are also adding to the pollution. In fact, it is estimated that 77 per cent of marine pollution originates from the land.

In particular, plastic waste is a major problem for the ocean.

Box 8.3 Sources of Marine Pollution

The main sources of marine pollution are:

- Land-based activities: More than 75 per cent of marine pollution originates from land. This includes the dumping of hazardous waste (including nuclear material), heavy discharges from industry and urban sewage systems, and inflow of fertilizers and pesticides from agricultural fields.
- Pollution from ships: About 9,000 tankers crisscross the ocean carrying goods, oil, and chemicals. They cause pollution through regular cleaning processes, accidental oil spills, transfer of oil between ships at sea, illegal discharges, emissions from engines, and so on.
- Pollution from ocean exploration: Off-shore drilling for oil and gas causes pollution through the dumping of waste from oil rigs, oil spills from accidents to rigs, and air pollution from burning of oil and gas. The exploration of the deep seabed for mineral as well as biological resources has negative consequences too.
- Marine debris: This includes sunken ships (with their cargo of hazardous material), lost fishing gear, etc.
- Ship breaking: Toxic material removed from old ships that are dismantled in ship-breaking yards (such as Alang in Gujarat).

Threats to Marine Living Resources

Marine living resources are under threat from human activities:

- Overfishing: Due to overfishing by large trawlers and factory ships, world fisheries have now collapsed and many fish species have disappeared forever.
- Bottom trawling: The trawlers that scrape the bottom destroy the organisms on the deep seabed.
- Invasion by alien species: Thousands of species are carried every day to new places by ballast water (used by ships to keep in balance). The alien species multiply rapidly in the new environment and destroy the local environment.
- Bycatch: Along with the targeted species, millions of sharks, dolphins, seabirds, and other species are caught in the nets and often thrown away.

Box 8.4 The Great Pacific Garbage Patch

It is the largest landfill in the world and it is floating on the Pacific Ocean. In 1997, Charles Moore, an American oceanographer, was sailing on the Pacific and found himself surrounded by rubbish. He was thousands of miles from land and yet he was moving through garbage for a whole week! That was how Moore discovered the Great Pacific Garbage Patch (GPGP).

There are in fact two large patches filled with more than 100 million tonnes of garbage, mostly plastic. The patches cover an area twice the size of the US. Garbage from land and ships tends to collect in some places due to swirling underwater currents.

Plastic forms 90 per cent of all trash floating in the world's oceans. UNEP (United Nations Environment Programme) estimated in 2006 that every sq. km of ocean contained over 18,000 pieces of floating plastic. Of the more than 90 billion kg of plastic the world produces each year, about 10 per cent ends up in the ocean. About 70 per cent of that eventually sinks, damaging life on the ocean floor, and the rest floats. Much of it ends up in garbage patches.

The plastic does not biodegrade, but just breaks into smaller and smaller pieces, called nurdles. Marine animals eat these nurdles, which can poison them or lead to deadly blockages. Nurdles also soak up toxic chemicals. As a result, the chemicals or poisons become highly concentrated. These poison-filled masses threaten the entire food chain. What goes into the ocean goes into marine animals and on to our dinner plates.

Albatrosses are the worst affected. On Midway Island, which comes into contact with parts of the Eastern Garbage Patch (EGP), albatrosses give birth to 5,00,000 chicks every year. Two hundred thousand of them die, many of them by consuming plastic fed to them by their parents, who confuse it for food. In total, more than a million birds and marine animals die each year from consuming or becoming caught in plastic and other debris.

Trawling the ocean for all of its trash is simply impossible and would harm plankton and other marine life. Meanwhile, the garbage patch is becoming larger. Unless we reduce the use of disposable plastics, the patch would double in size over the next decade.

- Unsustainable hunting: For example, several whale species are endangered due to massive hunting. Though there is a ban on whale-hunting, countries like Japan and Norway continue the practice.
- Shipping and coastal development: For example, thousands of Olive Ridley turtles that land every year on an Orissa beach to lay eggs are killed by propeller injuries, toxic contaminants in water, consuming debris like plastic bags, etc.
- Disease, destruction of habitats, pollution, and so on: Due to such reasons, many species are declining in numbers.

CORAL REEFS

Coral reefs are one of the natural wonders of the ocean. They are found in the shallow coastal zones of tropical and sub-tropical oceans. Apart from being aesthetically appealing, they are also among the world's oldest, most diverse, and most productive ecosystems.

Corals are formed by huge colonies of tiny organisms called polyps. They secrete calcium carbonate (limestone) to form a protective crust around their soft bodies. When they die, their outer skeletons remain as a platform for others to continue building the coral. The intricate crevices and holes in the coral catacombs become the home for 25 per cent of all marine species.

The colour of the coral comes from zooxanthellae, the tiny single-celled algae that live inside the tissues of the polyps. In return for the home provided by the polyps, zooxanthellae produce food and oxygen through photosynthesis.

Importance of Coral Reefs

Coral reefs are complex ecosystems that perform many ecological services. When polyps form their shells, they absorb some carbon dioxide as part of the carbon cycle. These reefs help protect the coastal zone from the impact of waves and storms. They act as nurseries for hundreds of marine organisms. In their biodiversity and intricacy of relationships, they can be called the rainforests of the ocean.

On the economic side, they provide fish, shellfish, building materials, medicines, and employment to people. Through tourism, they bring in valuable foreign exchange and give us the pleasure of enjoying the world under the sea.

Coral reefs are now in trouble everywhere (Box 8.5). They are very vulnerable to damage because they grow very slowly, get disrupted easily, and are very sensitive to variations in temperature and salinity.

Of more than 100 countries with large coral formations, 90 per cent are experiencing the decline of the reefs. South East Asia, with the most species of all coral reefs, is the most threatened region.

Box 8.5 Threats to Coral Reefs

Coral reefs around the world are being killed by a combination of reasons:

- Land-based pollution from agriculture, industrial and human waste.
- Collection of corals and shells for sale.
- Ballast water from ships.
- Dredging.
- Coastal development.
- Crown-of-thorns starfish which feed on the living coral.
- Destructive fishing practices like blast fishing and cyanide fishing.

When a reef becomes stressed, it expels the zooxanthellae, loses its colour and food, and ultimately dies. Such coral bleaching can occur even with an increase of one degree in water temperature.

Conservation of Coral Reefs

Some of the coral conservation measures are:

- Reserves: 300 coral reefs in 65 countries are protected as reserves and 600 more are under consideration.
- Creation of awareness: Convincing people not to buy fish that have been caught by destroying coral reefs and encouraging tourists not to destroy corals.
- Reducing pollution: Campaigning for the reduction of pollution of the ocean from land-based activities.
- Lobbying with governments for stricter regulations and policies for conserving corals.

RECENT REPORTS SAY OUR CORAL REEFS ARE IN CRITICAL DANGER!
WARNING
SEWERAGE OUTLET.

Gulf of Mannar
PIZZA

MANGROVES

Mangroves are unique salt-tolerant trees with interlacing roots that grow in shallow marine sediments. Often they are found just inland of coral reefs. Mangroves provide valuable ecosystem services. Their roots are the breeding grounds and nurseries for many fish species like shrimp and sea trout. The branches are nesting sites for birds like pelicans, spoonbills, and egrets. They stabilize the soil, preventing erosion, and provide protection to the coast during cyclones. They are more effective than concrete barriers in absorbing wave action.

Threats to Mangroves

Mangroves are under threat due to natural causes and human activities. Between 1985 and 2000, the world lost half of its mangroves. The natural causes include climate change, cyclones, trampling by wildlife, grazing, and damage by insects, oysters, and crabs.

Human activities that damage mangroves include the following:

- Indiscriminate tree felling and lopping, mainly for fuel wood, fodder, and timber.
- Indiscriminate conversion of mangroves for aquaculture, agriculture, mining, human habitation, and industrial purposes.
- Encroachments on publicly owned mangrove forest lands.
- Illegal large-scale collection of mangrove fruits for use in production of medicine.
- Discharge of industrial pollutants into creeks, rivers, and estuaries.
- Traditional use of dragnets in fishing, which often hampers regeneration of mangroves because young seedlings become entangled in the nets and get uprooted.

CORAL REEFS AND MANGROVES IN INDIA

Despite India's large size and long coastline, we have only a few coral reefs. These are found in the Gulf of Kutch, Gulf of Mannar, Lakshadweep, and the Andaman and Nicobar Islands, covering about 2,300 sq. km. They are all being damaged and destroyed at an increasingly alarming rate.

Mangroves occur in the Sunderbans, Andaman and Nicobar Islands, Gulf of Kutch, Orissa, Tamil Nadu, and Goa. They cover about 6,700 sq. km, which is about 7 per cent of the world's total mangrove area. During the last century, India lost 40 per cent of its mangroves.

Attempts are now being made to conserve mangroves and coral reefs in India.

Box 8.6 Conservation of Coral Reefs and Mangroves in India

Coral reef areas in India are protected areas under the Wildlife (Protection) Act 1972. Hence, they come under the forest department. Coral mining is prohibited under the Coastal Regulation Zone Rules.

With the support of the Global Coral Reef Monitoring Network (GCRMN), Indian scientists are monitoring the health of coral reefs, creating awareness, providing training, and promoting sound management practices. The Ministry of Environment and Forests (MoEF) has also set up a National Mangrove and Coral Reef Committee (NMCRC) to support research projects.

Measures for the conservation of mangroves include natural regeneration, afforestation (planting of mangroves), and protection (including conservation, setting up parks and reserves, and so on)

FISHERIES IN CRISIS

The world's fisheries are in fact in deep crisis. An environmental and social catastrophe is in the making, but most consumers of fish are unaware of the problem.

A massive increase in global fishing began in the 1950s and 1960s with the rapid induction of new technology: factory trawlers, satellite positioning, acoustic fish finders, spotter planes, huge nets, and so on. Huge trawlers began operating in all the seas catching all varieties of fish stocks. Soon the rate of harvest exceeded the rate of fish population growth.

The first sign of problem was the collapse of the world's largest fishery, the Peruvian anchovy in 1972. The decline in the north Atlantic fisheries started in the mid-1970s and over the next two decades most of the cod stocks in New England and eastern Canada collapsed. Centuries of fishing tradition came to an end.

The global fish catch increased five-fold between 1950 and 1980. The capacity of the fishing fleet, however, expanded twice as fast as the rise in catch. Now there are too many boats chasing too few fish.

We are now fishing down the food web. Once the larger fishes are exhausted, the fleets start catching the smaller fish. These being often the prey of the larger fish, there is further decline of the latter. Meanwhile, we are also removing the species living on the ocean floor through bottom trawling, which is like clear cutting a forest. Apart from overexploitation, fisheries are threatened by pollution of water bodies, climate change, destruction of mangroves and coral reefs.

For each species of fish, there is a maximum sustainable yield (MSY) that we could harvest annually leaving enough breeding stock for the population to renew itself. MSY is the amount we can catch every year indefinitely. We have already exceeded the MSY in the case of many important species. More than 80 per cent of the marine fish stocks are fully exploited, overexploited, or in a state of depletion. Large-scale fisheries are very likely to collapse by 2050.

COASTAL ZONE

The coastal zone is partly land and partly ocean. Thanks to the movement of tides, the coastal zone is an ever-changing ecosystem, rich in biodiversity. About 90 per cent of all marine species are found in this zone.

People tend to migrate to the coastal zone from inland areas. Apart from harbours, most of the large cities of the world are on the coast. The coastal zone is also the preferred place for tourism, hotels, resorts, and industries. Due to such diverse human activities, the environment of the coastal zone has been seriously affected:

- Mangroves that protect the coast have been destroyed.
- Coral reefs have been exploited and damaged.
- Fisheries are affected since fishers are displaced or denied access to the ocean.
- Tourism and other activities lead to heavy pollution on the beaches and this enters the ocean.
- Harbours often cause sea erosion on one side and sand accretion on the other side.

CONSERVING THE OCEAN AND THE COASTAL ZONE

The open oceans and their resources can be conserved only through international cooperation. Since 1972, a number international agreements and programmes have focused on controlling marine and coastal pollution:

- London Dumping Convention: Prevention of pollution by regulating the dumping of waste materials into the oceans.
- Basel Convention: Control of the movement of hazardous wastes across international borders.
- International Convention on the Prevention of Pollution from Ships (MARPOL): Prevention of pollution by ships from operational or accidental causes.
- Global Programme of Action for the Protection of the Marine Environment from Land-based Activities (GPA-LBA): Control of contaminants like sewage,

POPs, radioactive substances, heavy metals, oils, nutrients, sediments, and litter that enter the oceans from land.

A number of agreements cover the sustainable management of marine living resources. The main ones are:

- The Code of Conduct for Responsible Fisheries (CCRF), initiated by the UN Food and Agricultural Organization (FAO), to promote sustainable fishing practices.
- The UN Fish Stocks Agreement covers fish species that migrate over long distances.
- Convention on Biological Diversity (CBD) and the Convention on International Trade in Endangered Animal and Plant Species (CITES) cover marine species.

Countries have their own laws and rules to manage fisheries. For example, India has regulations on the use of trawlers and small boats in fishing. Coastal states often ban fishing during the spawning season of the fishes to conserve stocks.

Marine Protected Areas

Marine Protected Areas (MPAs) are like forest reserves. They are set up by countries to protect marine ecosystems, natural processes, habitats, and species. They provide the following benefits:

- Maintaining biodiversity and providing secure areas for species.
- Protecting habitats from damage by human activities and allowing damaged areas to recover.
- Providing areas where fish are able to spawn and grow to their adult size.
- Increasing fish catches in surrounding fishing grounds.
- Helping to support local communities that are linked to the marine environment.
- Serving as benchmarks for undisturbed, natural ecosystems, which can be used to measure the impact of human activities in other areas.

Though there are a large number of MPAs in the world, they do not cover all the important marine ecosystems. Many MPAs do not serve their purpose due to lack of money and other reasons.

India's MPAs cover about 5,000 sq. km, which is a small fraction of India's EEZ. The main MPAs are in the Gulf of Kutch, the Gulf of Mannar, and the Andamans.

Ocean Governance and the Law of the Sea

For a long time, the ocean area beyond the territorial limits of countries was governed by the concept of the 'freedom of the high seas'. This meant that any country or anyone could exploit the wealth of the oceans without any control.

The UN Convention on the Law of the Sea (UNCLOS), which came into force in November 1994, is a landmark agreement. Apart from setting the territorial limit as 12 nautical miles, this convention establishes for every country an exclusive economic zone (EEZ) up to a distance of 200 nautical miles from the shoreline. (One nautical mile is equal to 1.85 km.)

Every country has the exclusive right to exploit the resources of its EEZ. What is beyond the EEZ is the 'international area of the seabed', controlled by the UN. India has a coastline of over 7,000 km and an EEZ of 2 million sq. km, which is two-thirds of our land area.

According to the convention, the oceans have to be used only for peaceful purposes and should be managed in the interests of all, including future generations.

Protection of the Coastal Zone in India

India's Coastal Regulation Zone (CRZ) Notification of 1991 prohibits harmful activities such as the following in specified areas of the coastal zone:

- Setting up of new industries and expansion of industries.
- Manufacture and handling of hazardous substances.
- Land reclamation, bunding, or disturbing the natural formation of the coastline.
- Mining of lands, rocks, and other sub-strata materials.
- Large-scale extraction of groundwater.
- Construction activities in ecologically sensitive areas.

You can also take action to conserve the oceans and coastal zones.

Think about it: During the 30 minutes it took for you to read this chapter, 300 sea birds were killed by plastic waste.

Box 8.7 What You Can Do to Save the Ocean and the Coast

Taking the actions suggested in the other chapters will have a positive effect on the oceans and the coasts too. There are also specific things you can do to protect the oceans:

- Do not eat any fish species that are endangered.
- Do not take a vacation on a cruise liner. The cruise industry is a great polluter of the oceans.
- Boycott shows involving dolphins and other marine animals.
- Avoid cosmetics and jewellery made from the organs of marine animals.

If you live in a coastal area:

- Take part in local stream, river and beach cleanups—or organize one yourself. They may not solve the bigger problem, but the people will realize the adverse impact of their actions and habits.
- Join groups that protect sea turtles, which hatch on the shore.
- Check whether sewage or garbage in your area is pumped directly into the sea. Be conscious of this and any other potential sources of marine litter in your area. Demand that these are eliminated.

Box 8.8 Find Out and Be Surprised!

1. What percentage of all life on earth is found in the oceans?
2. How much of the ocean space has been explored by us?

Answer in *Appendix C*

Box 8.9 Hopeful Signs: Chilika Lake

Chilika Lake in coastal Orissa is the largest brackish water lagoon in Asia. Chilika had been facing a number of ecological and social problems. Heavy siltation was reducing the lake area by 2 sq. km every year. The salinity level was decreasing and many fish species were affected. Since 2000, however, the Chilika Development Authority has taken many steps to restore the lake with the involvement of the local people. The lake's environmental condition has improved considerably.

Box 8.10 Once by the Ocean

The shattered water made a misty din.
Great waves looked over others coming in,
And thought of doing something to the shore
That water never did to land before.
The clouds were low and hairy in the skies,
Like locks blown forward in the gleam of eyes.
You could not tell, and yet it looked as if
The shore was lucky in being backed by cliff,
The cliff in being backed by continent;
It looked as if a night of dark intent
Was coming, and not only a night, an age.
Someone had better be prepared for rage.
There would be more than ocean-water broken
Before God's last Put out the Light was spoken.

—Robert Frost*

*(1874–1963); American poet.

END PIECE

A quotation from the Canadian humorist, essayist, teacher, political economist, and historian Stephen B. Leacock (1869-1944):

It is to be observed that 'angling' is the name given to fishing by people who can't fish.

EXPLORE FURTHER

Books and Articles

Clover, Charles, 2008, *The End of the Line: How Overfishing Is Changing the World and What We Eat*, Berkeley: University of California Press.

Earle, Sylvia, 2009, *The World Is Blue: How Our Fate and the Ocean's Are One*, Washington, D.C: National Geographic Society.

Ellis, Richard, 2004, *The Empty Ocean: Plundering the World's Marine Life*, Washington, D.C: Shearwater Books.

Menon, Parvathi, 2003, 'A Conflict on the Waves', *Frontline,* Vol. 20, No. 6 (28 March), pp. 65-73 (Gulf of Mannar).

Pattnaik, A.K., 2003, 'Chilika: Blue Lagoon Once More', *The Hindu Survey of Environment 2003*, Chennai: The Hindu, pp. 147-53.

Pauly, Daniel, 2004, 'Empty Nets', *Alternatives Journal,* Vol. 30, No. 2 (Spring), pp. 8-10.

Pauly, Daniel and Reg Watson, 2003, 'Counting the Last Fish', *Scientific American* (July), pp. 42-7.

Safina, Carl, 1999, *Song for the Blue Ocean: Encounters Along The World's Coasts And Beneath The Seas*, New York: Owl Books.

Websites

Chilika Lake agitation against shrimp farms: www.angelfire.com/in/aquiline/chilika.htm
Chilika Lake: www.chilika.com
Coral reefs—UN World Conservation Monitoring Center: www.unep-wcmc.org/marine/data/coral_mangrove/index.html
Coral reefs—WWF: www.panda.org
Cousteau Society: www.cousteau.org
Great Pacific Garbage Patch: www.greatgarbagepatch.org
Gulf of Mannar Biosphere Reserve Trust www.gombrtrust.org/
Gulf of Mannar: http://envfor.nic.in/icrmn/dist/mannar.html
International Ocean Institute: www.ioinst.org
Marine Protected Areas: http://depts.washington.edu/mpanews/
Ocean Conservation Portal and Search Engine: www.oceanconserve.org
Sea Shepherd Conservation Society: www.seashepherd.org
UN Department of Ocean Affairs and the Law of the Sea: www.un.org/Depts/los/index.htm

9 A WORLD OF PEOPLE
Growing Population and Urbanization

USA Today *has come out with a new survey:*
Apparently three out of four people make up 75 per cent of the population.

—David Letterman*

THE STORY OF DHARAVI: DEPRIVATION AND DESPERATION, ENERGY AND ENTERPRISE!

Dharavi is a slum spread over 175 ha in the centre of Mumbai. Within the congested city, Dharavi has the highest density of population, an unbelievable 45,000 persons per ha. The residents come from many parts of India, driven to the city from their villages by drought, discrimination, or deprivation. There are potters from Kumbharvada in Saurashtra, *dhobi*s from Gujarat, and tanners from Tamil Nadu. Many began as workers and ended up as owners of small factories. Many of the old timers even live in two-storey houses. All of them, however, are 'illegal' occupants!

Living in Dharavi is not easy. Everywhere there are open drains, piles of uncleared garbage, filth, and pitiful shacks. Water supply and sanitation were once non-existent, now poor. The people suffer incredible hardships during the monsoon, with flooded lanes and rivers of sewage.

There are around 5,000 small industries in the area with a total turnover of Rs 20 billion. The industries include plastic recycling, garment-making, printing,

* (b. 1947); American television host and comedian.

zari making, leather products, and pottery. Leather products made here are exported to France and Germany. Every day, about 200 tons of snacks are produced in over 1,000 units.

When the migrants were banished to Dharavi years ago, it was a swamp outside the city. Now, it finds itself in the heart of the vastly expanded city in a strategic location between the two main railway lines. And so the land has become extremely valuable!

Over the years, unsuccessful attempts were made 'to develop' Dharavi. In early 2004, however, the state government announced a Rs 56 billion project to completely transform Dharavi. A Dharavi Development Authority (DDA) has been set up. According to the original plans, Dharavi would be divided into sectors, allocating space for residential, commercial, industrial, and recreational use. Every resident, who had come there before 1 January 2000 would be eligible for a free home of 270 sq. ft. The space released would be given to builders for commercial development.

The residents and NGOs have opposed the project. The original plans are being revised, but serious doubts about the idea itself still remain. Will the scheme work at all? What will happen to the large number of residents not covered by the scheme? Where is the space to build transit camps for at least 6,00,000 people during the construction phase? Will new slums be created in the process of redevelopment? What will happen to the thousands of illegal, polluting, but flourishing businesses?

The Dharavi Redevelopment Project had not taken off even in 2010, but we have wait for further news.

WHAT DOES THE STORY OF DHARAVI TELL US?

Managing the urban population is becoming a bigger and more complex problem every day. In particular, how do we treat the poor, who arrive in the cities by thousands each day? More than a billion people (a third of the urban dwellers) live in slums. Can we provide jobs, food, shelter, water, and sanitation for all of them? Such questions are likely to become more pressing in the coming decades.

POPULATION EXPLOSION

About 10,000 years ago, there were about five million people on earth, all hunter-gatherers. Then came agriculture and settlements, the population started growing, and by the fourteenth century the figure had reached 500 million. Thereafter it remained steady for about 500 years, until the scientific and industrial revolutions took place in Europe.

Between 1850 and 1950, the world population doubled to a figure of 2 billion.

Box 9.1 What You Should Know about Population and Urbanization

- The world's population has been rapidly increasing, though there are now signs of slower growth.
- Most of the population growth now occurs in the developing countries.
- By the middle of this century, India is very likely to overtake China in population. India's efforts in family planning have not produced the expected results.
- Half the world's population now lives in cities.
- By 2025, India's three largest metros will be among the ten largest cities of the world.
- The unceasing migration from villages into cities and the lack of space and housing for the urban poor leads to slums like Dharavi.
- Population growth and urbanization place great pressures on the natural resources.
- Innovative and transparent approaches like the ones used in Surat and Curitiba can mitigate urban problems.

By that time, exponential growth had set in and the five billion mark was reached in 1987, and the six billion mark in 1999.

We now add one billion to the population every 12 years or so. The world population is projected to cross 7 billion in early 2011 and reach about 8.1 billion by 2025. Such growth is perhaps beyond the earth's capacity to support.

POPULATION DISTRIBUTION IN THE WORLD

The land area covers only 30 per cent of the earth's surface and, of that area, 80 per cent is not conducive to human settlement. This includes deserts, the polar regions, tropical rainforests, tundra, and the like.

Most of the population lives in coastal areas, river basins and cities. At least 40 per cent of the world's population lives within 100 km of the coast. By 2025, this figure is likely to double. Again, many of the megacities of the world are on the coast.

There is a stark difference in the population growth pattern between the developing and the industrialized nations. Nearly 99 per cent of all population increase now takes place in developing countries, while the population size is static or declining in the industrialized nations. Among the major industrialized nations, only the US has significant population growth, mainly because of immigration.

The low birth rates in industrialized countries are beginning to challenge the health and financial security of their senior citizens. On the other hand, the developing countries add over 80 million to the population every year and the poorest of those countries add 20 million. This worsens poverty levels and threatens the environment.

Table 9.1 The Ten Most Populous Countries: 2010 Status and 2050 Estimates

2010 Status			2050 Estimates		
Rank	Country	Population (millions)	Rank	Country	Population (millions)
1	China	1,338	1	India	1,748
2	India	1,189	2	China	1,437
3	US	310	3	US	423
4	Indonesia	235	4	Pakistan	335
5	Brazil	193	5	Nigeria	326
6	Pakistan	185	6	Indonesia	309
7	Bangladesh	164	7	Bangladesh	222
8	Nigeria	158	8	Brazil	215
9	Russia	142	9	Ethiopia	174
10	Japan	127	10	Congo	166

Source: Population Reference Bureau, 2010 Report.

Table 9.1 lists the ten most populous countries and shows how the list is likely to change between 2010 and 2050. Of the three industrialized countries in the list now, only the US is expected to remain in the 2050 list.

Note that by 2050, India is expected to be well ahead of China. Obviously, China is doing a better job of controlling its population. The populations of Pakistan and Bangladesh are likely to double by 2050. Thus, China and the Indian sub-continent together could be home to about 40 per cent of the world's population.

GROWTH PATTERN OF INDIA'S POPULATION

There have been four phases in the growth of the population in India:

- 1901–21: Stagnant population (238 to 251 million)
- 1921–51: Steady growth (251 to 361 million)
- 1951–81: Rapid high growth (361 to 683 million)
- 1981–2001: High growth, with some signs of slowing down (683 to 1027 million)

Some features of India's population (according to the 2001 census) are:

- Almost 40 per cent of Indians are younger than 15 years of age.
- About 70 per cent of the people live in more than 5,50,000 villages.
- The remaining 30 per cent live in more than 200 towns and cities.

We have about 16 per cent of the world's population, but only 2.3 per cent of the land and 1.7 per cent of forests. As we will see in Chapter 10, Indian agriculture faces problems like soil degradation, water scarcity, and decreasing biodiversity.

INDIA'S RESPONSE TO POPULATION GROWTH

India was the first country in the world to start a family planning programme. It was launched in 1952, when our population was about 400 million. Fifty years and many programmes later, we have more than one billion people!

Poverty, low literacy and education levels among women, lack of consistent support from the government, poor planning, and bureaucratic inefficiency are some of the reasons why the family planning programme has not been a big success. Poverty drives many people to have more children so that there will be more working members in the family. The desire to have male children is also a factor in increasing the family size.

URBANIZATION AND GROWTH OF LARGE CITIES

For the poor rural inhabitant facing problems of drought, discrimination, and deprivation, the city offers what the person desperately needs—employment. Most of the world's economic activities take place in cities and even a person without any skills can almost always find something to do in a city. The city also offers the hope of a better life with comforts not available in most rural areas.

The UN Report on World Urbanization Prospects (2009 Revision) provides the following key findings:

- With more than half the population living in cities, the world has now become more urban rather than rural. China, India, and the US account for

36 per cent of the world urban population. Globally, the level of urbanization is expected to rise from 50 per cent in 2009 to 69 per cent in 2050.

- The world urban population is expected to increase by 84 per cent by 2050, from 3.4 billion in 2009 to 6.3 billion in 2050. Virtually all of the expected growth in the world population will be concentrated in the urban areas of the less developed regions,
- Twenty-one megacities (population more than 10 million each) accounted for 4.7 per cent of the population in 2009. That is, just about one in every twenty people on earth lived in megacities. The number of megacities is projected to increase to 29 in 2025, at which time they are expected to account for 10.3 per cent of the world urban population.
- Tokyo, the capital of Japan, is today the most populous urban agglomeration with 36.5 million people. The next largest urban agglomerations are Delhi with 22 million inhabitants, São Paulo in Brazil and Mumbai, each with 20 million inhabitants.
- In 2009, Asia was home to about half of the urban population in the world. Over the next four decades, Asia will experience a marked increase in its urban populations. By mid-century, most of the urban population of the world will be concentrated in Asia (54 per cent) and Africa (20 per cent).

Table 9.2 lists the ten largest cities in 1950 and 2009. It also gives the expected list in 2025.

We can see the rapid growth of the cities of the developing world between 1950 and 2009. Note that the three largest metros of India are in the 2025 list.

Table 9.2 Ten Largest Cities of the World

Rank	City	1950 Pop. (million)	City	2009 Pop. (million)	City	2025 Pop. (million) (Estimate)
1	New York	12.3	Tokyo	36.5	Tokyo	37.1
2	London	8.7	Delhi	21.7	Delhi	28.6
3	Tokyo	6.9	Sao Paulo	20.0	Mumbai	25.8
4	Paris	5.4	Mumbai	19.7	Sao Paulo	21.7
5	Moscow	5.4	Mexico City	19.3	Dhaka	20.9
6	Shanghai	5.3	New York	19.3	Mexico City	20.7
7	Essen	5.3	Shanghai	16.3	New York	20.6
8	Buenos Aires	5.0	Kolkata	15.3	Kolkata	20.1
9	Chicago	4.9	Dhaka	14.3	Shanghai	20.0
10	Calcutta	4.4	Buenos Aires	13.0	Karachi	18.7

Source: UN World Urbanization Prospects—2009 Revision.

ENVIRONMENTAL IMPLICATIONS OF THE POPULATION GROWTH AND URBANIZATION

Population growth and urbanization will place greater pressures on the natural resources, but there are environment friendly alternative measures that would mitigate the problems:

- Land: Expanding cities will encroach into the surrounding areas, converting any kind of land, including fertile fields, into new colonies. Land will also be needed for infrastructure like roads, highways, industries, tourist facilities, and educational complexes. Better land-use planning, creation of satellite towns, and similar measures are necessary.
- Food: Our ability to produce enough food for all will be tested. Unless there is a second Green Revolution, which does not demand high inputs as the first one did, it will be difficult to feed the billions.
- Forests: We can expect increasing encroachment of forest areas and exploitation of forest resources. It will become increasingly difficult to protect reserved forests and national parks. The only way is to make the local people partners in conservation.
- Water supply and sanitation: Water scarcity will become more severe and sanitation targets will not be met unless water conservation measures and ecological sanitation are aggressively implemented.
- Energy resources: As our population and consumption grow, our energy demand will soar. Fuel wood and fossil fuels will have problems of scarcity and higher prices, unless wood plantations are encouraged and renewable energy sources are promoted.
- Cities and civic services: As cities and slums grow, water scarcity will intensify, more and more waste will pile up, the air quality will drop, public transport will be overloaded, traffic jams will increase, and so on. Cities must be planned better.
- Housing: The shortage of housing, which is already a problem, will become severe in the future. There is a need to implement mass housing using the principles of ecological architecture.

While there are many horror stories of urban situations, there are also hopeful ones. Examples are the planning of Curitiba in Brazil (Box 9.2) and the post-plague transformation of Surat in Gujarat (Box 9.3).

Just to let you know: During the time it took for you to read this chapter, about 2,500 more people were born on this planet!

WATER TANK
WORLD POPULATION DAY !

DHARAVI

Box 9.2 Hopeful Signs: The City of Curitiba

Would you like to live in a city set amidst greenery, where the streets are clean, the air is breathable, the public transportation is excellent, there are no traffic jams, the citizens are happy and are fully involved in running the city? Then, head for Curitiba in Brazil!

Curitiba's transformation into an ecocity (2.5 million population) is due to the ideas and efforts of Jaime Lerner, an architect who became the city's mayor. Lerner set out to establish a municipal government that would find simple, innovative, and inexpensive solutions to the city's problems. He wanted an accountable, transparent, and honest government. It should be ready to take risks and also correct itself. It should be, above all, an environment friendly city.

Lerner introduced innovations in many aspects of the city's working:

- Greenery: He planted 1.5 million trees and created a series of interconnected parks crisscrossed with bicycle paths.
- Transportation: The city has perhaps the world's best public transport, with clean and efficient buses carrying 1.9 million passengers every day at low cost. The buses run on high-speed dedicated lanes and the bus stops are connected to the bicycle paths. The population has doubled since 1974, but the traffic has reduced by 33 per cent.
- City planning: High-density development is restricted to areas along the bus routes, where low-income groups are housed. Every high-rise block includes two floors of shops, reducing the need to travel far for shopping. The city centre is a large pedestrian zone connected to the bus stations, parks, and bicycle paths.
- Welfare: Slums do exist, but each poor family can build its own home. A family gets a plot of land, building materials, two trees and a one-hour consultation with an architect. The poor receive free medical aid and childcare and infant mortality has reduced by 60 per cent since 1977. There are 40 feeding centres for street children.
- Waste management: There is a labour-intensive garbage purchase programme. 7,00,000 poor people collect and deliver garbage-filled bags in exchange for bus tokens, surplus food, or school notebooks. The city recycles 70 per cent of paper and 60 per cent of metal, glass, and plastic. The recovered material is sold to the city's 500 units in an industrial park.
- Industries: The city developed an industrial park with services, housing, and schools, so that workers can cycle or walk to work. There are strict pollution control laws.
- Education: All the children study ecology, while adults have special environmental courses. The older children are given training and apprenticeship in environment-related areas like forestry, water pollution, and ecorestoration.

All this has happened even as the population grew from 3,00,000 in 1950 to 2.2 million in 1999, as rural people flocked to the city. The city expects to grow by another million by 2020.

Box 9.3 Hopeful Signs: The City of Surat

In 1994, the city of Surat in Gujarat (with a population of two million) was struck by pneumonic plague. During the heavy monsoon that year, there was flooding and waterlogging in low-lying areas. Hundreds of animals died. In September, the whole system gave way, inviting the plague.

Surat was known for its diamond business and also for its filth. Despite being one of the richest civic bodies in the country, the Surat Municipal Corporation (SMC) had failed to provide basic sanitation and clean drinking water to a majority of the city's population.

When the plague struck, Surat became a shunned city and 60 per cent of the citizens fled. The economy was devastated and export of food grains from Surat was banned. Gradually, however, the plague subsided and normalcy began to return. The strange fact was that the plague did not change matters much. A miracle was to happen, however.

In May 1995, S.R. Rao, IAS, took over as the municipal commissioner of Surat. Faced with a city traumatized by the plague, Rao set to work. When he left Surat two years later, it was ranked as the second cleanest city in India, after Chandigarh.

This is what Rao did:

- Formed a team to work together on policy and implementation and delegated powers down the line.
- Launched a 'Surat First' programme to involve citizens, companies, and institutions in working for the city's welfare.
- Built water and sanitation facilities in the slums and improved the living conditions of the sanitation workers.
- Demolished illegal structures, even those built by powerful people.
- Enforced cleanliness in all eating places.
- Set up a night cleansing system: Every street was scrubbed at night and garbage bins cleared so that the people awoke to a clean city each morning.
- Established 275 surveillance centres for monitoring public health.

The level of sanitation improved from 35 to 95 per cent, while solid waste removal increased from 40 to 97 per cent. Disease rate went down by 70 per cent.

Rao and his officers went to the filthiest of slums. The initial scepticism of the citizens gave place to a spirit of cooperation, when they saw that the corporation meant business.

The most important thing Rao did was to put a system in place. He made the civic staff realize how a proper use of their powers would earn them respect. Later, it became a matter of prestige for the staff to continue the work that was started after the plague.

Box 9.4 Find Out and Be Surprised!

1. What percentage of the world's population lives in the Northern Hemisphere?
2. During the twentieth century, the demand for one human need increased faster than the population growth. Which one?

Answers in *Appendix C*

Box 9.5 Blessing: A Poem on Dharavi

The skin cracks like a pod.
There never is enough water.

Imagine the drip of it,
the small splash, echo
in a tin mug,
the voice of a kindly god.

Sometimes, the sudden rush
of fortune. The municipal pipe bursts,
silver crashes to the ground
and the flow has found
a roar of tongues. From the huts,
a congregation: every man woman
child for streets around
butts in, with pots,
brass, copper, aluminium,
plastic buckets,
frantic hands,

and naked children
screaming in the liquid sun,
their highlights polished to perfection,
flashing light,
as the blessing sings
over their small bones.

—Imtiaz Dharker*

*Poet, artist, and filmmaker; Dharker was born in Pakistan but now lives in India.

END PIECE

George Bernard Shaw (1856-1950), Irish playwright and socialist, had a very positive view of the people in this world:

The world is populated in the main by people who should not exist.

EXPLORE FURTHER

Books and Articles

Bunsha, Dionne, 2004, 'Developing Doubts', *Frontline* (18 June), pp. 44-5.
Jamwal, Nidhi, 2004, 'Dharavi Dreaming', *The Telegraph* (22 February).
McKibben, Bill, 2007, *Hope, Human and Wild: True Stories of Living Lightly on the Earth*, Minneapolis, Minnesota: Milkweed Editions. (Includes a chapter on Curitiba.)
Sharma, Kalpana, 2000, *Rediscovering Dharavi*, New Delhi: Penguin India.

Websites

Curitiba: www.context.org/ICLIB/IC39/Meadows.htm
www.dismantle.org/curitiba.htm
www.infra.kth.se/courses/1H1142/Curitiba_Paper_1.pdf
http://www.sustainable.org/casestudies/international/INTL_af_curitiba.html
Dharavi: www.dharavi.org

Films

Blessed by the Plague (on Surat, Arunabh Bhattacharjee, 1999)
Dharavi—A Slum for Sale (Lutz Konermann, 2010)

10 A HUNGRY WORLD
Land, Soil, Food, and Agriculture

Farming looks mighty easy when your plough is a pencil,
and you're a thousand miles from the cornfield.

—DWIGHT D. EISENHOWER*

THE STORY OF PUNJAB: DEGRADING SOIL, DEPLETING WATER

Punjab is called the granary of India and its farmers are known for their hard work and for quickly adopting modern technologies. Today, however, Punjab is a story of degraded soil, depleted water tables, reduced productivity, and farmer suicides. How did things change?

The Green Revolution arrived in Punjab in the mid-sixties. As we will see later in this chapter, the Green Revolution involved the use of special seeds, chemical fertilizers, and pesticides. In Punjab, the productivity of land increased and the farmers' income rose. The farmers began to increase the area under wheat and rice cultivation.

As cultivation became more intensive and the gap between crops narrowed, the farmers felt the need to buy tractors and other machinery. As their cash income grew, more machinery and chemicals followed.

Over 40 years, the area under rice cultivation increased from 6 to 60 per cent and the area under tubewell irrigation went up by 80 per cent. There are 14 lakh tubewells taking out groundwater all the time.

* (1890–1969); 34th President of USA.

The result is that water tables have fallen drastically all over Punjab. In 90 per cent of the area, water is only available at a depth of 10 m or more. In many places, it is down to 20 m. This has also created acute scarcity of drinking water in villages. Now 11,800 habitations out of a total of 12,400 are facing drinking water problems. The groundwater may be exhausted by 2025.

Punjab buys 18 per cent of the total pesticides used in the country, while it has only 2.5 per cent of the agricultural area! No wonder, pesticides are found in cereals, milk, butter, fruits, vegetables, and even in infant formula. What is worse, they are present in people's blood too. Many areas report serious health problems like cancer, early ageing, and skeletal fluorosis.

With so much of pesticide and fertilizer residues, Punjab's soil has been declared sick. Over the years, the excessive use of chemicals has degraded the soil by killing many of the beneficial microorganisms. The farmers have to keep on adding fertilizers. Naturally, the cost of production has been going up.

With costs increasing all the time, the debt burden on the farmers of Punjab has reached about Rs 22,000 crore. The average debt per farmer is about Rs 50,000. More than 13,000 farmers have committed suicide in recent years.

Many others are migrating to cities. The situation is so desperate that the residents of Kamalwala near Fazilka have put up their village 'for sale'.

WHAT CAN WE LEARN FROM THE STORY OF PUNJAB?

We now live in a world of degrading land, depleting soil, farmer suicides, and hungry people.

Nature has a way of hitting back, when we try to exploit its resources in an unsustainable way. That is what happened with the Green Revolution in Punjab and elsewhere. In the mad rush for higher productivity, farmers adopted multicropping with huge inputs of fertilizers, pesticides, and water. Such overexploitation could not be sustained.

FOOD AND AGRICULTURE

Food comes from three sources:

- Croplands that provide 76 per cent of the total, mostly grains.
- Range lands that produce meat mostly from grazing livestock accounting for about 17 per cent of the total food.
- Fisheries that supply the remaining 7 per cent.

In this chapter, we focus on croplands and related agriculture.

Food and agriculture are complex subjects since they are influenced by a variety of factors including:

Box 10.1 What You Should Know about Land, Soil, Agriculture, and Food

- We now live in a world of degrading land, expanding deserts, depleting soil, farmer suicides, and hungry people.
- There is enough food available in the world to feed everyone, but poverty keeps millions of people hungry.
- While the Green Revolution increased crop production dramatically, it has had a negative impact on biodiversity, soil health, water resources, land, food, and human health.
- Per capita food production has been going down and food prices have been rising sharply.
- The wise option for agriculture is to shift gradually to sustainable methods such as organic farming, minimizing or altogether avoiding chemical fertilizers, and chemical pesticides.
- There are many examples of successful adoption of sustainable farming.

- Environmental factors (land, soil, water, plant species and seeds, energy needs, climate change, natural disasters, and so on).
- National and international policies (procurement from farmers, storage, trade regulations, patenting of species, credit for farmers, and others).
- Science and technology issues (fertilizers, pesticides, genetic modification, and so on).

All the problems and issues discussed in the chapters of this section with regard to water, energy, biodiversity, pollution, and population together have an impact on agriculture and food.

The current major concerns in the world about food and agriculture are the following:

- Feeding the ever-increasing population.
- Soaring prices of food items.
- Plight of small farmers and the unviability of farming.
- Impact of environmental factors (including climate change) on food production.
- Contamination of food by hazardous substances.
- Role of transnational corporations (TNCs) in controlling the production and trade in seeds, fertilizers, pesticides, and grains.

GLOBAL FOOD CRISIS

Global food production was steadily rising from 1990 to 2006. Since then, however, per capita food output has been declining in the least-developed countries. Global reserve stocks of food have been falling since the 1980s, and more rapidly since 2000. The stock-to-use ratio is now the lowest since 1960s. Global stocks of wheat and rice would probably last just for 12 weeks.

There is enough food in the world to provide the minimum requirement for all. Yet, more than one billion people do not have enough to eat. The problem is that the hungry are too poor to buy the available food. The paradox is this: The very farmers who grow our food are often so poor that they do not get enough to eat! (See also Box 10.2.)

After reaching a peak during the 1974 oil crisis, food prices had been steadily coming down. Since 2007, however, prices have been increasing sharply in many parts of the world. The main reasons for this price rise seem to be:

- Failed harvest in large producer-countries due to drought, floods, or climate change.
- Diversion of grains and land for rearing animals due to high demand for meat and milk.
- Diversion of land for growing plants for making biofuels to meet the energy shortage.

Box 10.2 Hunger Facts

- Between 1980 and 1995, some progress was made in reducing chronic hunger in the world. Since then, however, hunger has been slowly but steadily on the rise. Today, every sixth person in the world goes hungry.
- Over 65 per cent of the world's hungry live in only seven countries: India, China, the Democratic Republic of Congo, Bangladesh, Indonesia, Pakistan, and Ethiopia. India has the largest number of hungry people—about 230 million.
- More than 60 per cent of chronically hungry people are women. Every year, hunger kills 12 million children worldwide.
- Even as the number of hungry people increases, we lose over 50 per cent of the edible crop harvest through wastages at various stages of the supply chain.

- High input costs due to high oil prices (since fertilizers and pesticides are mostly made from petroleum).
- Low reserve stocks.
- Speculation and criminal fixing of prices.

Most of the world's small farmers are suffering due to the following reasons:

- Decreasing soil productivity.
- Water scarcity.
- Uncertain weather.
- Low prices of their produce.
- Increasing inputs of fertilizers and pesticides.
- Vagaries of government policies.

As a result, small farmers fall into the debt trap. Many have been selling their lands and migrating to cities for employment. Many others have committed suicide, often consuming the very pesticide they were using on their land.

LAND AS A NATURAL RESOURCE

The land area of the earth, about 140 million sq. km, occupies less than a third of the surface. Yet, it is vital for our existence since it is land that:

- Preserves terrestrial biodiversity and the genetic pool.
- Regulates the water and carbon cycles.
- Acts as the store of basic resources like groundwater, minerals, and fossil fuels.
- Becomes a dump for solid and liquid waste.
- Forms the basis for human settlements and transport activities.

Even more importantly, the topsoil, just a few centimetres in thickness, supports all plant growth and is hence the life support system for all organisms, including human beings.

Box 10.3 Why Indian Farmers are Committing Suicide

According to India's National Crime Records Bureau (NCRB), about 2,00,000 farmers committed suicide between 1997 and 2008. In most cases, an intolerable debt burden drove them to take their lives. They had borrowed money for the inputs in their farming—seeds, fertilizers, and pesticides. The Green Revolution had increased the yields, but also created problems like soil degradation, falling productivity over time, increasing use of fertilizers, greater proneness to diseases, and excessive withdrawal of groundwater. These factors, combined with high costs, low prices, uncertain weather, and lack of easy credit make farming uneconomical.

The small farmer is now in the clutches of a new breed of entrepreneurs. They provide all the inputs at high prices, give credit at very high interest rates, and buy the crops at low prices. A crop failure due to spurious or low quality seeds, pest attack, water scarcity, or drought is often the last straw. The farmer is perpetually in debt and often the only escape is through suicide. The Government of India had initiated a loan waiver programme in 2008, but that covered only loans from banks, not moneylenders.

The farmer has no health cover for his family and any major expense only drives him further into debt. When the farmer commits suicide, his wife and children are in deeper trouble.

Condition of the World's Land Surface

The biggest change that we have made to the ecosystems of the world is not by building cities, industries, and highways, but converting them into agricultural fields. Then, through our intensive agriculture and other practices, we have also been degrading the land surface everywhere.

UN studies estimate that 24 per cent of all usable land (excluding mountains and deserts, for example) has been degraded to such an extent that its productivity is affected. The main causes are agricultural mismanagement (planting unsuitable crops, poor crop rotation, poor soil and water management, excessive input of chemicals, frequent use of heavy machinery like tractors, and so on), deforestation, fuel wood consumption, overgrazing, establishment of industries, and urbanization.

Soil erosion and degradation, which occur due to loss of green cover, strong winds, chemical pollution, and others, have severe effects on the environment. They affect the soil's ability to act as a buffer and filter for pollutants, regulator of water and nitrogen cycles, and habitat for biodiversity.

Waterlogging, salinization, increasing desertification, and impact of mining are some of the other problems affecting land. Intensive surface mining of land for metals and minerals destroys all vegetation in the area and pollutes the landscape with the dust that is thrown up. Once the available material is mined out, large craters are left behind. When hills that act as watersheds are mined away, the water tables go down. The processing of the mined material, often done on site, using in many cases mercury, cyanide, and large quantities of water, pollutes rivers and other water bodies. The enormous amount of waste material like slag is left behind as dangerous heaps.

IMPORTANCE OF SOIL

Life on earth depends on a very thin skin of soil that covers land. Trees, plants, shrubs, herbs and grasses—all depend on this layer for their sustenance. Soil performs many vital functions for the environment and for us:

- Plants grow on soil and give us food.
- Soil maintains the biodiversity of earth.
- Plants and soil support each other.
- Soil cleans water as it percolates down.
- Soil reduces flooding by absorbing the water.
- By retaining water during the monsoon and releasing it slowly later, soil prevents drought.
- Soil regulates the balance of moisture, oxygen and heat in our atmosphere, to create climate and maintain the stability of our weather.
- Soil also absorbs harmful gases like carbon dioxide and methane.

Role of Microorganisms and Worms in Soil

Microorganisms and earthworms play an important role in building a healthy soil:

- Microorganisms in soil convert organic matter into humus. The dark humus enables the soil to perform functions such as water retention, good drainage, aeration, and protection against erosion.
- The organisms also make the soil fertile so that plants can grow well. They release minerals from the sub-soil and provide nutrients to the plant roots.
- Earthworms constantly build rich topsoil. They take in decaying organic matter and excrete castings, which make good fertilizer.

State of the World's Soil

Most soils on earth are becoming depleted and unhealthy in the following ways:

- Erosion of topsoil: The rich topsoil is carried away by flowing water and wind.
- Salinization of soil: Excess salts in irrigation water remain in the soil and affect productivity.
- Lack of nutrients: This can happen when little or no fertilizer is added to the soil, as in the case of Africa.
- Contamination of soil: This is the result of excessive use of chemical fertilizers and pesticides, as in the case of India and other Asian countries.

GREEN REVOLUTION

India now produces annually 180–220 million tons of food grains, 5 million tons of meat products, and 6 million tons of fish. The increase in grain production since the 1960s is ascribed to the Green Revolution.

Until about the middle of the 20th century, farmers in India needed very little input from outside. They grew a variety of crops and kept aside some part of the harvest as seeds for the next year. They used organic manure and natural pesticides. Even if one crop failed, there were others to save the farmer from ruin.

Life was not easy, but the farmers were not in deep crisis either. However, India's population was growing and food was being imported. This was also the problem with many poor countries of the world.

The solution came in the form of the Green Revolution with the promise of plenty. The key element was new seeds called high yielding varieties (HYVs). Developed first in Mexico and then taken to many countries, the new seeds increased the yield per hectare to very high levels.

The Green Revolution came as a package: HYVs along with high inputs of chemical fertilizer, pesticide, water, and agricultural machinery like tractors. It was an energy-intensive method: apart from the energy that went into the making of the inputs, energy was again needed to run the machinery and to pump water.

PESTICIDE COS
HERE, MY
A MUCH
DEAL FOR
YOU DON'T USE CHEMICALS, DO YOU?
ORGANIC VEGETABLES

W's

The farmers also had to buy new seeds, fertilizers, and pesticides every year. The Green Revolution encouraged farmers to plant the HYVs replacing the indigenous ones. Over the years, many traditional species have disappeared. In place of 30,000 varieties of rice, Indian farmers now predominantly plant just 12 species.

Initially, the seeds came from the National Seeds Corporation (NSC) or other certified government agencies. From about 1998, however, international seed companies entered the Indian market. They need mass markets and they aggressively promote the use of a small number of varieties.

The farmers found that, year after year, the inputs had to be increased to maintain the productivity levels. The new varieties were more prone to diseases and pest attacks. The soil has also been getting degraded and drained of its nutrients through the excessive use of chemicals. (See the Punjab story above.)

GLOBAL FOOD SYSTEM

In the global food system, there are a large number of farmers in the fields. At the other end, there are a large number of consumers. In between, there are governments, moneylenders, traders, and large TNCs. Together, they control the main activities:

- Determine what the farmer should plant.
- Supply seeds, fertilizers, and pesticides.
- Provide formal and non-formal credit to farmers.
- Procure the raw material and process it where possible.
- Organize the supply chain through wholesalers and retailers.

Today TNCs control 40 per cent of the world trade in food and this proportion is increasing with globalization and liberalization. They tend to offer low prices to farmers and charge the consumers high. They lobby with governments to bend rules or to change them. They even fix prices.

Our current system of agriculture and food makes unsustainable demands on energy, water, and soil. It causes harm to people and communities involved both in production and consumption.

WAY OUT OF THE CRISIS IN AGRICULTURE AND FOOD

Some of the solutions to the crisis are:

- Adopt methods of sustainable agriculture.
- Discourage the diversion of land for biofuels.
- Reduce the use of cereals and food fish as animal feed.
- Regulate the activities of TNCs that control the supply and prices of agricultural inputs and food items.

Box 10.4 Biotechnology and Genetically Modified Crops

Biotechnology is the manipulation of living organisms or cells to create a product or an effect. Biotechnology makes use of microorganisms, such as bacteria, or any biological substance, for specific purposes.

An important aspect of biotechnology is genetic engineering or genetic modification (GM). All life is composed of cells that contain genes, and genes are made of molecules called DNA (deoxyribonucleic acid). The DNA molecule of an organism contains information about its characteristics and behaviour. Genetic engineering manipulates the genes in an organism to change its characteristics. It can move a favourable gene from one organism to another.

Genetic modification can make a plant resistant to specific pests or diseases. Bt cotton and Bt brinjal are examples. Bt stands for the bacterium *Bacillus thuringiensis*, which acts as a microbial pesticide. Each strain of this bacterium specifically kills one or a few related species of insect larvae. Scientists introduce the Bt gene into the plant's own genetic material. Then the plant manufactures the substance that destroys the pest.

Genetic modification can also produce new varieties of plants with some desired characteristics, such as herbicide tolerance, virus tolerance, and resistance to drought and salinity. However, there are strong opinions in favour of and against the introduction of genetically modified crops. Those who support genetically modified crops give arguments such as:

- Genetically modified is the only way to produce enough to feed the increasing world population.
- Plants will be resistant to pests, droughts, water salinity, and even climate change.
- There will be a reduction in the use of fertilizers and pesticides.

Those who oppose GM raise questions such as:

- How do we know the unintended consequences of genetically modified foods, such as long-term effects on humans and other organisms?
- We know that a genetically modified crop contaminates the non-genetically modified varieties grown nearby. What would be the effects of such contamination on biodiversity?
- Can the farmer grow seeds for the next planting or is he perpetually dependent on the seed company? Would GM lead to domination of agriculture by a few large companies?
- Is it ethical and safe to introduce animal genes into plants and vice versa?

- Support the small farmer through easy credit, affordable inputs, and simple technology

Biotechnology and GM crops have been hailed as wonder solutions to the food crisis, but there are issues concerning them.

Sustainable Agriculture

Sustainable agriculture refers to our ability to produce food indefinitely, without causing irreversible damage to ecosystem health. Sustainable agriculture has biophysical, socio-economic, and environmental aspects:

- The impact of various agricultural practices on soil properties and processes should not affect crop productivity in the long term.
- The farmers should be able to obtain the necessary inputs, manage the resources, and make a decent living in the long term.
- Agriculture should use natural resources like water and land in such a way that these resources remain available indefinitely.

Organic Farming

A way of moving towards that goal is to make a gradual shift from chemical agriculture to organic farming, which is based on the following principles:

- Nature is the best role model for farming since it uses neither chemicals nor poisons and does not demand excessive water.
- Soil is a living system and not an inert bowl for dumping chemicals.
- Soil's living populations of microbes and other organisms are significant contributors to its fertility on a sustained basis and must be protected and nurtured at all costs.
- The total environment of the soil, from soil structure to soil cover, is more important than any nutrients we may wish to pump into it.

Organic farming practices include the following:

- Avoidance of artificial fertilizers and pesticides: Organic farming uses natural fertilizer (like compost and green manure), biopesticide, and biological pest control.
- Feeding the soil: In organic farming, we add a lot of green matter to the soil.
- Maintaining the fertility of soil: Organic farming avoids the overexploitation of the soil and keeps the soil healthy.
- Avoidance of monocropping: Planting a single species over a piece of land makes the plants vulnerable to pest attacks. Organic farming promotes mixed crops that benefit one another.

Organic farming that blends traditional knowledge with modern scientific ideas can, over a period of time, reverse soil degradation and improve soil health.

Box 10.5 *Hopeful Signs: Organic Farming*

ORGANIC FARMING IN CUBA

The small island of Cuba had for long depended on the Soviet Union for exchanging its sugar for fertilizers, oil, and grains. When the Soviet Union collapsed in 1989, Cuba's food supplies were seriously threatened. The people were facing starvation.

The Cuban government's answer was a major shift to organic farming. Today, thousands of gardens across Cuba produce organic vegetables and many other crops. Organic farming on small family plots currently provides employment to 3,25,000 people out of a population of 12 million. The Cuban success in shifting to organic farming points the way to the rest of the world.

ORGANIC FARMING IN INDIA

Ramesh Chandar Dagar in Sonipat, Haryana, makes more than Rs 1 million a year from 1 ha. of land through integrated organic farming. His method includes bee-keeping, dairy management, biogas production, water harvesting, and composting. The key element in Dagar's farm is the cyclic, zero-waste approach. The paddy waste goes into vermicomposting as well as mushroom production, the dung is fed into the gobar gas plant, the sludge from which goes into the composting pit, and so on.

Dagar has set up the Haryana Kisan Welfare Club (HKWC) and its 5,000 members are now busy spreading the message of integrated organic farming. Led by farmers like Dagar, Nammalwar in Tamil Nadu, and Narayan Reddy in Karnataka, organic farming is spreading to many parts of India.

Conversion to organic farming will mean initial problems and economic losses as the soil recovers, but it will be a better option in the long term. The success of Cuba in shifting to organic farming is a classic case.

There are more than 1,000 registered organic farmers in India and there are many more who are practising organic methods. There are organic farmers who get as much (if not more) yield in comparison to chemical farming.

You can also take steps to make the food system better and save yourself.

Box 10.6 What You Can Do to Save the Food System and Protect Yourself

- Examine every item that you eat: Where is it from? Who made it? When? What processing has it gone through?
- Examine your food habits: Reduce or avoid meat.
- Keep away from all forms of fast food and processed food.
- Beware of advertisements that manipulate your food instincts.
- Eat locally and seasonally, not just 'organic'.
- Keep away from supermarkets.
- Plant a kitchen garden.
- Organize a network of consumers and organic farmers and arrange direct supply from farms to homes.

Box 10.7 Find Out and Be Surprised!

1. There are more than one billion hungry people in the world. What is the number of obese people and why are they obese?
2. What percentage of the world's cereal production is used as animal feed?

Answer in *Appendix C*

Box 10.8 The Song of Right and Wrong

Feast on wine or fast on water,
And your honor shall stand sure
If an angel out of heaven
Brings you something else to drink,
Thank him for his kind attentions,
Go and pour it down the sink.

*—G.K. Chesterton**

*(1874–1936); English writer.

Think about it: During the 30 minutes it may have taken for you to read this chapter, USD 3 million worth of food was thrown away in American homes.

END PIECE

Alfred E. Newman is the fictional mascot and iconic cover boy of *Mad* magazine. Along with his face, the magazine also includes a short humorous quotation credited to Neuman with every issue's table of contents. Here is an example:

We are living in a world today where lemonade is made from artificial flavors and furniture polish is made from real lemons.

EXPLORE FURTHER

Books

Kukathas, Uma (ed.), 2009, *The Global Food Crisis*, Current Controversies Series, Detroit: Cengage.

Patel, Raj, 2008, *Stuffed & Starved: What lies behind the world food crisis*, New Delhi: HarperCollins India.

UNEP, 2009, *The Environmental Food Crisis—The Environment's Role in Averting Future Food Crisis*, a UNEP Rapid Response Assessment, Nairobi: UNEP.

Websites

Greenpeace India: www.greenpeace.org/india
Genetically engineered as modified seeds:
www.indiatogether.org/agriculture/gegm.htm
Organic farming in India: www.ofai.org
Organic farming in India: www.indiatogether.org/agriculture/organic.htm
Green Revolution in Punjab:
www.livingheritage.org/green-revolution.htm
Organic farming opportunities: www.wwoof.org

11 A WORLD PAST THE TIPPING POINT?

When written in Chinese,
the word crisis is composed of two characters.
One represents danger and
the other represents opportunity.

—JOHN F. KENNEDY*

THE STORY OF THE AMAZON RAINFOREST: AN ENVIRONMENTAL TIPPING POINT?

The Amazon river basin in South America contains the largest rainforest on earth. The basin covers about 40 per cent of South America and includes parts of eight countries: Brazil, Bolivia, Peru, Ecuador, Colombia, Venezuela, Guyana, and Surinam.

The Amazon rainforest, also known as Amazonia, is one of the world's greatest natural resources and covers about 8 million sq. km. Its vegetation continuously recycles carbon dioxide into oxygen, producing about 20 per cent of the earth's oxygen. Hence, it is called the 'Lungs of our Planet'.

The Amazon rainforest is the drainage basin for the Amazon river and its many tributaries. The northern half of the South American continent is shaped like a shallow dish. All the rain that falls in the river basin drains into Amazon rainforest. Fifty per cent of the 2.7 m. of annual rainfall returns to the atmosphere through the foliage of trees.

* (1917–1963); 35th President of United States.

Amazonia contains nearly 40,000 plant species, over 500 mammals, over 475 reptile species, and about 30 million insect types. It also contains the world's richest diversity of birds, freshwater fish, and butterflies. It is also one of the world's last refuges for jaguars, harpy eagles, and pink dolphins.

The Amazon rainforest is a massive sink that soaks up around 1.5 billion tons of carbon annually. Thus it is a critically important buffer against global warming and helps in stabilizing local and global climate.

Human activities have wrought havoc on the Amazon rainforest. Large areas of the rainforest have already been destroyed and, by 2030, 55 per cent could be gone. The land is being cleared for cattle ranches, mining operations, logging, and subsistence agriculture. Some forests are being burned to make charcoal to power industrial plants.

The combination of rampant deforestation, fires, and rising temperatures could devastate the rainforest ecosystem within 50–60 years. The Amazon rainforest could then shrink to one-third of its original size. Such a contraction would result in countless extinctions, losses in vital freshwater sources, a decline in regional rainfall, and the weakening of one of the world's most important carbon sinks.

Since the Amazon rainforest creates at least half the rainfall needed to sustain itself, the loss of forest cover and general drying would create a feedback cycle and large areas of the forest would revert to savanna. This process would release billions of tons of additional carbon into the atmosphere and worsen global warming.

Most likely due to global climate change, the Amazon rainforest experienced severe droughts in 2005 and 2010. Massive death of trees resulted, releasing more than 10 billion tons of carbon. This has raised fears that the vast forest is on the verge of a tipping point, where it will stop absorbing greenhouse gas emissions and instead increase them. If the Amazon switches from a carbon sink to a carbon source that prompts further droughts and mass tree deaths, such a feedback loop could cause runaway climate change, with disastrous consequences.

WHAT DO WE LEARN FROM THE STORY OF THE AMAZON RAINFOREST?

Global warming is seriously affecting sensitive and vital ecosystems like the Amazon rainforest and the Arctic. When forests die and ice melts, further warming results, more trees die and more ice melts. Where would this escalating cycle lead us?

INTERCONNECTIONS IN NATURE

We live in a world of complex interconnections and feedback loops. Every action that a human being or an animal takes has an effect on one part of the environment, which in turn has another effect on another part, and so on. Boxes 11.1, 11.2, and 11.3 give three stories of surprising interconnections.

Even small actions, repeated many times, may lead to major consequences. Suppose there is a small forest on a hill. A person living near the forest needs wood for building his house and for burning as fuel. He cuts down a tree,

Box 11.1 Connections: Drink Coffee in the US and Make the Songbird Vanish in South America!

How can drinking a cup of coffee in a chain shop in the US lead to the disappearance of songbirds in South America?

In recent years, chains of coffee shops have proliferated in the US and many other Western countries. The coffee comes from Central and South America and the growers now face a heavy demand. Traditionally coffee is grown under the shade of the rainforest (without using chemicals) and the ripe berries are handpicked. In order to increase productivity and meet the new demand, the growers have been shifting to unshaded plantations treated with fertilizers and pesticides. Over the past 25 years, about 50 per cent of the shade trees have been replaced with unshaded plantations.

The shade plantations have always provided shelter for a large number of birds, vertebrates, and insects. Many migratory songbirds spend the winter in the rain forests. With a decrease in shade trees, there has been an alarming decline in the population of birds like warblers and orchard orioles.

Conservation groups like the Rainforest Alliance and the American Birding Association (ABA) have initiated programmes to certify coffee as 'shade-grown' and encourage consumers to ask for such varieties. In fact, one brand already available is called Songbird Coffee!

Box 11.2 Connections: Get Rid of Malaria, but Invite the Plague!

In mid-twentieth century, malaria was rampant in the Indonesian island of Sabah, earlier known as North Borneo. In 1955, the WHO (World Health Organization) began spraying the island with the dieldrin (a chemical related to DDT) on the mosquitoes. The attempt was successful and malaria was almost eradicated.

The dieldrin, however, did other things too. It killed many other insects including flies and cockroaches. The lizards ate these insects and they died too. So did the cats that ate the lizards. Once the cats declined, rats proliferated in huge numbers and there was a threat of plague. WHO then dropped healthy cats on the island by parachute!

The dieldrin had also killed wasps and other insects that consumed a particular caterpillar, which was somehow not affected by the chemical. The caterpillars flourished and ate away all the leaves in the thatched roofs of the houses and the roofs started caving in.

Ultimately, the situation was brought under control. However, the unexpected chain of events showed the importance of asking at every stage the question: 'And then what?'

thinking that his small action cannot do any damage. However, everyone in the area does the same. Soon the forest is gone!

Without the forest to absorb the water, all the rainwater flows away down the hill. The springs and streams dry up and there is water scarcity in summer. The flowing water takes away the topsoil too and plants do not grow well. Food shortage and poverty are the result. Ultimately, rainfall too becomes less. Thus the cutting down of trees sets off a chain of events and ultimately, nature and the community suffer.

There are innumerable interconnections in nature. Often we cannot predict all the effects of our actions. Ecology and environmental science, however, tell us what is known about such connections and consequences. We should keep the information in mind when we take small or big actions.

Box 11.3 Connections: Weather Changes in Brazil, Forest Declines in Karnataka!

Brazil is the largest grower of coffee in the world, accounting for 30 per cent of the total production. However, droughts and frosts often destroy Brazil's coffee crops. The frequency of such attacks have increased over the last few decades and, in the mid-1990s, Brazil lost half of its output.

World coffee prices shot up and it was an opportunity for countries like India to increase their production for export. Three contiguous districts in south India—Kodagu in Karnataka, Nilgiris in Tamil Nadu, and Waynad in Kerala—account for 57 per cent of India's production. The growers here increased their plantation areas, but needed more manure.

The Indian growers prefer organic manure, which gives the coffee a distinct taste and value. They wanted huge quantities of dung, but did not want to use their valuable land to rear cattle. The dung had to come from outside.

The ideal source turned out to be a cluster of villages at the periphery of the Bandipur National Park in Karnataka. The farmers here were growing low-yielding millets and pulses. The expenses were high, but the income was low.

The coffee growers had somehow located the right place to get the dung: the agriculture was not profitable, cattle were abundant, and (as a bonus) free and unlimited fodder was available—from the Bandipur Park!

Soon, the villages became dung factories. Every day hundreds of cattle were taken into the forest for grazing and the dung was collected in the evening. Lorry loads of dung were sent to the coffee estates and the villagers made good money. A whole industry came into being with dung collectors, agents, lorry owners, and others.

Since the demand for dung was insatiable, the villagers bought more cattle. They bought the cattle cheap, since they did not need the milk-yielding varieties, but just any that 'could amble through the forests and defecate'. They did not even retain any dung for their own farming. With the money they earned from the dung, they bought subsidized fertilizers!

The forest periphery was now under a double threat. To the perennial fuel wood collection was now added heavy grazing by the cattle. Tree regeneration was affected, the park's wildlife had less forage, and degradation set in.

Meanwhile, Brazilian coffee started doing well again and Indian coffee industry went into a crisis. The dung trade, however, has acquired a steam of its own. The dung now goes to the ginger, chilly, and tea plantations of Kerala. When the demand goes down, the villagers sell the cattle for meat, again to Kerala.

The poor villagers of Bandipur have found a new livelihood—at the cost of the forest. Should we let them continue? Or should we ban the grazing and take away their incomes? Can we ever implement a ban on grazing? Hard questions with no easy answers!

STATE OF THE WORLD

The main conclusions from the last seven chapters are:

- Water is becoming scarce in many parts of the world.
- There is increasing energy shortage along with higher costs.
- It is an increasingly toxic world due to the huge amounts of waste we produce and the resultant pollution. This toxicity is affecting our health.
- We are rapidly losing our forests and many species, with unknown consequences.
- The great ocean, the inland seas, and the coasts are being exploited and polluted beyond measure.
- The world's population has been rapidly increasing, though there are now signs of slower growth.
- We now live in a world of degrading land, expanding deserts, depleting soil, farmer suicides, and hungry people.

The *State of the World 2010*, a report of the Worldwatch Institute, Washington, DC, had this to say about our consumption:

- In 2008 alone, people around the world purchased 68 million vehicles, 85 million refrigerators, 297 million computers, and 1.2 billion mobile phones.
- While the world population increased 2.2 times between 1960 and 2006, the consumption of goods and services went up 6 times.
- To meet this consumption, 60 billion tons of resources are now extracted annually—about 50 per cent more than just 30 years ago.
- This is why our ecological footprint (see Chapter 3) is now 1.4. In other words, people are using much more of earth's capacity than is available annually.

The key to this huge amount of production and consumption has been the availability of cheap energy from fossil fuels. The burning of fossil fuels has led to a carbon dioxide concentration in the atmosphere of 390 ppm (February 2010). If we go on with 'business as usual', the median global temperature is likely to go up by 4.5°C by 2100. Even if all countries meet their targets in reducing greenhouse gas emissions, temperatures would still go up by 3.5°C.

If that happens, it is very likely that ocean levels would increase by two or more metres due to the melting of Greenland or western Antarctica ice sheets. This in turn would cause massive coastal flooding and potentially submerge entire island nations. The one-sixth of the world that depends on glacier or snowmelt-fed rivers for water would face extreme water scarcity. Vast portions of the Amazon rainforest would become savannahs, coral reefs would die, and many of the world's most vulnerable fisheries would collapse. All of this would lead to major political and social disruptions—with environmental refugees projected to reach up to one billion by 2050.

HERE, MY CHILD, A MUCH BETTER DEAL FOR YOU!
PESTICIDE CO's
THE XYZ

CLIMATE CHANGE
OUT THERE WE WANT TO CREATE A PLAGUE!
I'M IN DANGER

If everyone lived like Americans, the planet could sustain only 1.4 billion people. That may be obvious, but how many people can the earth support, if everyone earns just what the citizens of Jordan and Thailand get on average? The surprising answer is 6.2 billion, while the current population is already 6.8 billion. This means that even relatively basic levels of consumption are no longer sustainable.

Box 11.4 How can Things Look so Good and yet be so Bad?

There are many reasons why outwardly everything seems normal, even though the environment is degrading rapidly:

- We do not pay the real costs of the resources we use. For example, till recent times, petroleum products were heavily subsidized by the government. The price of oil did not reflect the fact that reserves are limited and that burning of oil causes climate change resulting in heavy economic and social costs.
- We are using up nature's capital instead living off the annual interest.
- We do not see or hear what is happening, because the media does not report all the relevant, but unwelcome, news. The government too does not tell us the true story.
- When we are faced with fresh problems, we make adjustments, hoping that 'someone out there' will soon sort out the problem. We do not realize that some problems are permanent and will only get worse. For example, we started buying water in summer in recent years and now we do so throughout the year.

IS THE WORLD PAST THE TIPPING POINT?

All the data given above indicates that we are near or past the tipping point in many aspects of the environment. Oil production will soon peak or may have already peaked. This may also be true of the following:

- Grain production (total and per capita).
- Uranium production.
- Freshwater availability per capita.
- Arable land in agricultural production.
- Wild fish harvests.
- Extraction of some metals and minerals (including copper, platinum, silver, gold, and zinc).

The data is staring us in our faces and yet we continue with our preoccupation with economic progress and growth rates. We believe that environment is not such a serious problem (Box 11.4).

Of course, it is not the environment alone that causes concern. We have other issues such as increasing disparities, terrorism, increasing militarization, corruption, financial crises, and so on. But all these issues and the environment are interconnected.

The situation, however, is serious, but not hopeless. In the next part of the book we will find out how the world has been responding to the crisis, how we can find a cure for it, where we can draw hope from, and what we can do to help heal the planet.

END PIECE

We are hurtling towards disaster, but we do not want to accept it and take action. It is like the man who was falling from the fiftieth floor. Even as he reached the tenth floor, he said, 'So far, things are fine!'

EXPLORE FURTHER

Websites

Amazon Rainforest: en.wikipedia.org/wiki/Amazon_Rainforest
www.blueplanetbiomes.org/amazon.htm
www.mongabay.com
Worldwatch Institute: www.worldwatch.org

THE REMEDY

prologue

The Man who Planted Trees

I was then young and fond of travelling. Once I was walking in an area I had never visited. The land was dry and hot. I walked on for three days and I did not meet anyone. I did come across some villages, but these had been abandoned. The houses and even the temples were crumbling.

I realized I was in an area where the people used to make charcoal from wood. They went on cutting trees until there was nothing left. The land became barren and dry. The wells dried up and there was no water. The people had to leave the villages and go away.

As I walked on the hot and dusty land, I felt the howling wind that made things even worse. My water bottle was empty and I was very thirsty. Just then, I saw something like a tree at a distance. I walked towards it and, to my great relief, I saw a shepherd. He had some sheep and a dog.

The shepherd did not show any surprise on seeing me. He gave me some water and took me to his hut. The stone hut was very neat and tidy. I noticed that the man was also neatly dressed. He did not talk much. He just made me welcome. It was clear he expected me to spend the night as his guest.

He made some simple food and shared it with me. As I rested after the long walk, he was busy. He brought a bag full of seeds to the table. He examined carefully each seed and put aside the good ones. When he had selected a hundred seeds, he put them in a separate pile. We then went to sleep.

In the morning, he let his sheep out for grazing, with his dog to guard them. He set off with a stick and the

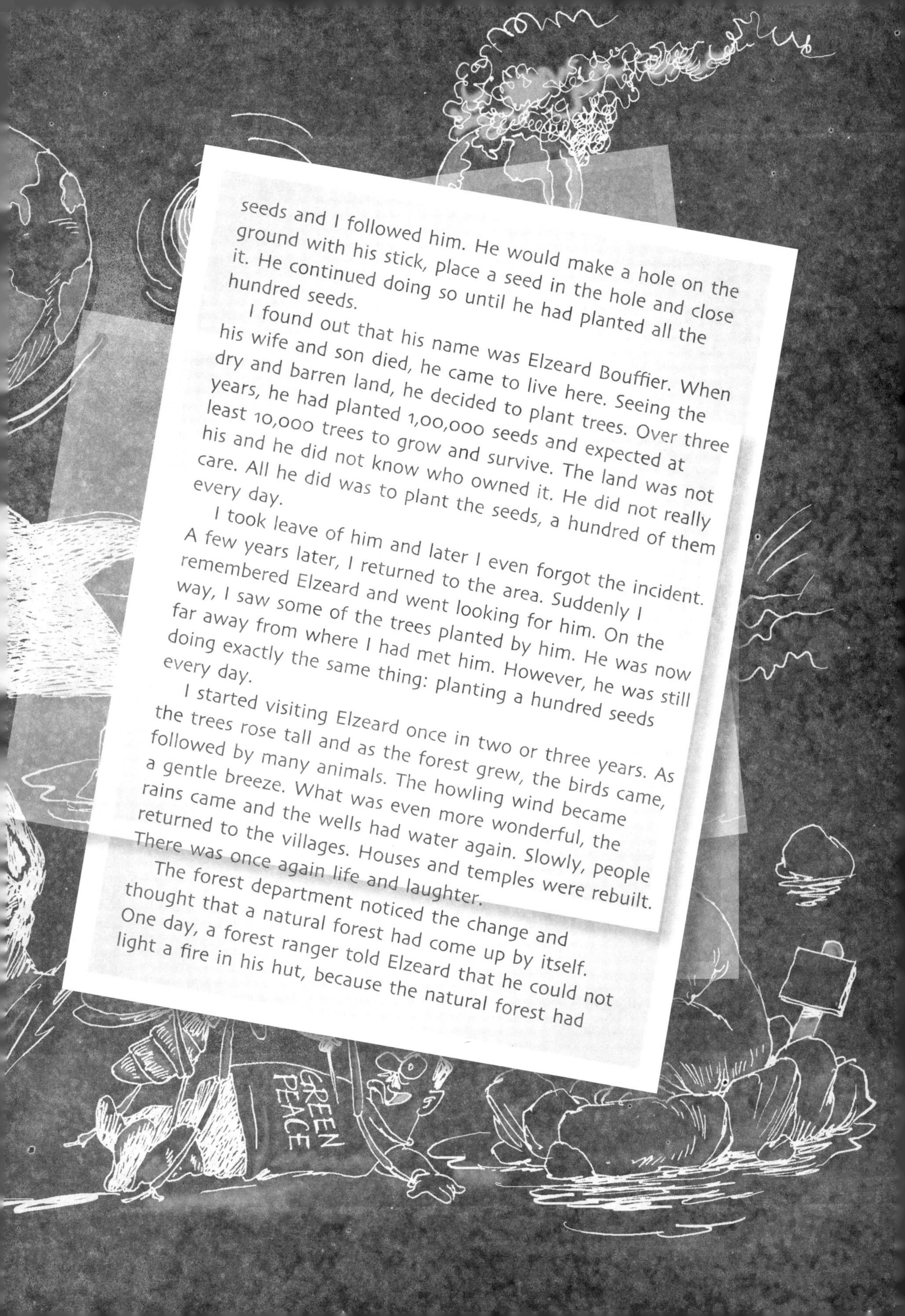
seeds and I followed him. He would make a hole on the ground with his stick, place a seed in the hole and close it. He continued doing so until he had planted all the hundred seeds.

I found out that his name was Elzeard Bouffier. When his wife and son died, he came to live here. Seeing the dry and barren land, he decided to plant trees. Over three years, he had planted 1,00,000 seeds and expected at least 10,000 trees to grow and survive. The land was not his and he did not know who owned it. He did not really care. All he did was to plant the seeds, a hundred of them every day.

I took leave of him and later I even forgot the incident. A few years later, I returned to the area. Suddenly I remembered Elzeard and went looking for him. On the way, I saw some of the trees planted by him. He was now far away from where I had met him. However, he was still doing exactly the same thing: planting a hundred seeds every day.

I started visiting Elzeard once in two or three years. As the trees rose tall and as the forest grew, the birds came, followed by many animals. The howling wind became a gentle breeze. What was even more wonderful, the rains came and the wells had water again. Slowly, people returned to the villages. Houses and temples were rebuilt. There was once again life and laughter.

The forest department noticed the change and thought that a natural forest had come up by itself. One day, a forest ranger told Elzeard that he could not light a fire in his hut, because the natural forest had

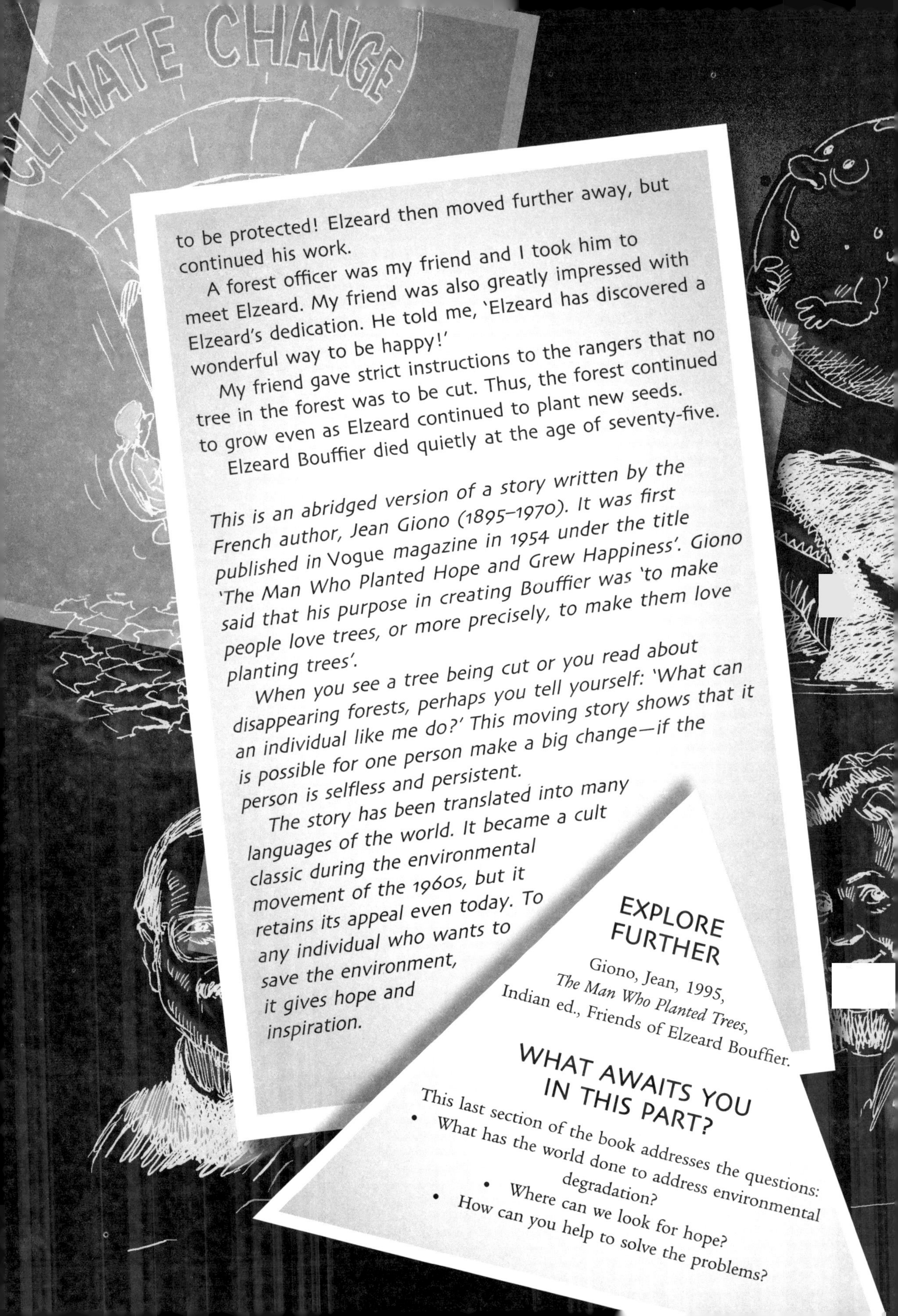

to be protected! Elzeard then moved further away, but continued his work.

A forest officer was my friend and I took him to meet Elzeard. My friend was also greatly impressed with Elzeard's dedication. He told me, 'Elzeard has discovered a wonderful way to be happy!'

My friend gave strict instructions to the rangers that no tree in the forest was to be cut. Thus, the forest continued to grow even as Elzeard continued to plant new seeds.

Elzeard Bouffier died quietly at the age of seventy-five.

This is an abridged version of a story written by the French author, Jean Giono (1895–1970). It was first published in Vogue magazine in 1954 under the title 'The Man Who Planted Hope and Grew Happiness'. Giono said that his purpose in creating Bouffier was 'to make people love trees, or more precisely, to make them love planting trees'.

When you see a tree being cut or you read about disappearing forests, perhaps you tell yourself: 'What can an individual like me do?' This moving story shows that it is possible for one person make a big change—if the person is selfless and persistent.

The story has been translated into many languages of the world. It became a cult classic during the environmental movement of the 1960s, but it retains its appeal even today. To any individual who wants to save the environment, it gives hope and inspiration.

EXPLORE FURTHER

Giono, Jean, 1995, *The Man Who Planted Trees*, Indian ed., Friends of Elzeard Bouffier.

WHAT AWAITS YOU IN THIS PART?

This last section of the book addresses the questions:

- What has the world done to address environmental degradation?
- Where can we look for hope?
- How can you help to solve the problems?

12 WHAT HAS THE WORLD DONE ABOUT THE CRISIS?

Why should I do anything for posterity?
What has posterity ever done for me?

—GROUCHO MARX*

THE STORY OF ANIL AGARWAL: THE ENVIRONMENTALIST WHO DID NOT GIVE UP

Anil Agarwal was a renowned environmentalist of India. He graduated from the prestigious Indian Institute of Technology, Kanpur (IIT-K) and, unlike many of his fellow students, chose to stay in India and become a science journalist. Attending the 1972 United Nations Conference on the Human Environment (Stockholm Conference) and covering the Chipko Movement (see Chapters 7 and 13) were turning points for him and he became an ardent environmentalist.

Deeply influenced by the ideas of Gandhiji, he realized the importance of the environment and natural resources for the poor people of India. He saw clearly that, if India were to focus only on its Gross National Product (GNP) and in the process destroys its 'Gross Nature Product', it would only lead to more poverty, loss of livelihoods, and greater unemployment.

Anil set up the Centre for Science and Environment (CSE) in Delhi and brought out a series of citizens' reports on the state of India's environment. These reports were prepared in cooperation with a number of individuals,

* (1890–1977); American comedian and film star.

voluntary organizations, grassroots activists, and others. In addition, CSE has been publishing books and other material on various environmental issues such as water, pollution, climate change, and health.

In 1992, Anil launched *Down To Earth*, a fortnightly newsmagazine on science and environment. The aim was to cover human aspirations, endeavours and struggles, global technologies, as well as the politics behind national and international policies and developments. The magazine has a wide readership including concerned citizens, NGOs, lawyers, teachers, students, industry leaders, government officials, researchers, and others. It remains the pre-eminent magazine of its kind in the country.

In India, Anil took up a variety of issues such as air pollution, green rating for industries, and environmental education. Thanks to the CSE campaign, Delhi's air became cleaner though the introduction of compressed natural gas (CNG) as the fuel for buses, trucks, and autorickshaws.

Anil also took up international environmental issues such as global warming and climate change. He argued on behalf of the poor and the disempowered in the global arena and wanted to make the industrialized countries accountable for their abuse of the environment.

Anil died at the young age of 54 due to cancer. He was convinced that changing environmental conditions, lifestyle, and consumption patterns were the cause of a majority of the new breed of deadly diseases like cancer. Even when he was sick, he began creating awareness about these issues and to bring policy changes.

Anil Agarwal spent the greater part of his life answering the question, 'How can India get more from our natural resource base and at the same time conserve the natural resources?' This remains the most important question even today.

WHAT DOES THE STORY OF ANIL AGARWAL TELL US?

Anil made us understand that environmental issues were very different in the industrialized countries and the poorer countries. Even within a country, ecosystems and societies vary widely and so do the environmental problems and their solutions. Anil also pointed out that only decentralized decision-making could match the enormous cultural diversity of Indian villages.

GLOBAL ENVIRONMENTAL MOVEMENT: THE BEGINNINGS

It was 1945, the end of the Second World War, and a devastated Europe. Though the US played a key role in winning the war, it did not experience the same kind of destruction. In fact, it was able to help Europe and Japan rebuild their economies.

Rapid industrialization of the Western countries followed and giant leaps were made in science, technology, agriculture, and medicine. The 'idea of progress' (see Chapter 3) was back with increased intensity. Large countries like India were becoming independent and it appeared that the world was moving towards progress and prosperity for all. Within 15 years, however, the environmental impact of rapid economic growth began to be felt.

Rachel Carson and *Silent Spring*

The first person to sound a warning was marine biologist Rachel Carson (see Prologue to Part I), whose 1962 book *Silent Spring* exposed the dangers of excessive use of the pesticide, DDT. The book met with stiff opposition, not just from the chemical industry. That was because it seemed to question the very idea of progress and economic development.

Development promised better employment prospects, more money in the pocket, greater equality, better health, freedom from want, and a more prosperous society. In fact, it promised the happiness that every individual was after.

Carson and others pointed out that development had, in fact, brought environmental devastation, health risks, inequality, poverty, and exploitation. They began to question the American dream and its high level of consumption. This was not acceptable to many.

In 1968, Paul Ehrlich wrote the book *The Population Bomb*, on the connection between rising human population, resource exploitation, and the environment. In 1972, the Club of Rome, a group of concerned intellectuals, published *Limits to Growth*, authored by Donella Meadows and others. It was based on a computer simulation of the world economy. It showed that unending economic development could only lead to the dwindling of natural resources. This study too was denounced as a false prediction of doomsday.

However, ordinary people had begun to take notice of environmental degradation. On 22 April 1970, the first Earth Day was observed and an estimated 20 million people participated in peaceful demonstrations across the US for air and water cleanup and preservation of nature.

In 1972, Stockholm hosted the UN Conference on the Human Environment (UNCHE). Prime Minister Indira Gandhi was the only outside head of state to attend it. At the conference there was a major disagreement between the developed north and the developing south.

The countries of the north wanted population control and environmental conservation everywhere. Those of the south pointed out that poverty was their main problem and this could only be solved through industrial development. They felt that the north, which had caused most of the environmental destruction through their industrialization, had no right to advise the south not to make any progress. At the conference, Indira Gandhi made the famous statement, 'Poverty is the worst polluter.' One positive result of the conference was, however, the setting up of the UN Environment Programme (UNEP).

Sustainable Development

In 1983, the UN set up the World Commission on Environment and Development (WCED) with Gro Harlem Brundtland of Norway as the chairperson. The 1987 WCED Report, called *Our Common Future*, emphasized the need for an integration of economic and ecological systems. The commission supported the concept of 'sustainable development' and defined it as: 'Development that meets the needs of the present without compromising the ability of future generations to meet their own needs.'

People welcomed the term 'sustainable development' and the concept of inter-generational responsibility. The definition, however, raised many questions: What are 'needs'? How can they be the same for everyone? Can development and environmental conservation go together at all? Such questions have never been clearly resolved.

Rio Summit and After

After Stockholm, the major effort was the United Nations Conference on Environment and Development (UNCED) held in 1992 in Rio de Janeiro. Attended by more than 100 Heads of State and 30,000 participants, UNCED approved several documents including:

- The Rio Declaration on Environment and Development listing 27 principles of sustainable development.
- *Agenda 21*, a detailed action plan for sustainable development in the twenty-first century.
- The convention on biological diversity.

Box 12.1 Environmental Timeline: Selected Events 1975–2009

- 1975: The Convention on International Trade in Endangered Species of Flora and Fauna (CITES) comes into force.
- 1980: The International Union for the Conservation of Nature (IUCN) releases its *World Conservation Strategy*. It identifies the main agents of habitat destruction as poverty, population pressure, social inequity, and trading regimes.
- 1982: The UN World Charter for Nature adopts the principle that every form of life is unique and should be respected regardless of its value to humankind. It calls for an understanding of our dependence on natural resources and the need to control our exploitation of them.
- 1987: Montreal Protocol on Substances that Deplete the Ozone Layer is adopted.
- 1994: UN Convention on the Law of the Sea (UNCLOS) comes into force. It establishes rules 1) concerning environmental standards and enforcement provisions dealing with marine pollution, and 2) the exploitation of the deep seabed.
- 2005: The Kyoto Protocol enters into force. It binds developed countries to goals for reductions in greenhouse gas emissions.
- 2005: UN releases the Millennium Ecosystem Assessment. More than 1300 experts from 95 countries provide scientific information concerning the consequences of ecosystem change for human well-being.
- 2009: Climate change negotiations open in Copenhagen with great expectations and the presence of world leaders. The outcomes, however, were vague. No concrete action was agreed upon.

CLIMATE CHANGE
REPENT!
THE END
IS NEAR
GLOBAL
WARMING
IS HERE!

CLIMATE CHANGE
DDT! DDT!
THE XYZ
GREEN PEACE

The next conference, World Summit on Sustainable Development, popularly known as Rio+10, was held in Johannesburg, South Africa, in 2002. This conference recognized that implementation of the Rio agreements had been poor. It marked a shift from agreements in principle to more modest but concrete plans of action.

International Agreements

Apart from the Stockholm, Rio de Janeiro, and Johannesburg meetings, there have been a number of international conferences, world commissions, conventions, and agreements on environmental issues.

However, the implementation of the agreements has been a mixed bag. Here are some examples:

- The decisions taken at the Rio summit were hardly implemented.
- CITES has been a qualified success.
- Montreal Protocol has been implemented well and the ozone layer is likely to stabilize during this century.
- UNCLOS was never ratified by some of the powerful countries like the US, but it has been accepted and acted on by many other countries.
- The Kyoto Protocol was not followed by most countries.

As the environmental movement expanded in the twentieth century, several international organizations (such as Greenpeace, World Wide Fund For Nature [WWF], World Conservation Union, and Friends of the Earth) were also set up. They conduct studies, publish reports, conduct campaigns, and lobby with governments.

Environmental Conservation in India

The Constitution (Forty-Second Amendment) Act of 1976 explicitly incorporates environmental protection and improvement. Article 48A, which was added to the directive principles of state policy, declares: 'The state shall endeavour to protect and improve the environment and to safeguard the forests and wildlife of the country.'

Article 51A(g) in a new chapter entitled 'Fundamental Duties', imposes a similar responsibility on every citizen 'to protect and improve the natural environment including forests, lakes, rivers and wild life, and to have compassion for living creatures.'

In addition, Article 21 of the constitution states: 'No person shall be deprived of his life or personal liberty except according to procedure established by law.' This Article protects the right to life as a fundamental right. The courts have interpreted this Article to mean that the enjoyment of life, including the right to live with human dignity, encompasses within its ambit the protection and preservation of environment.

In 1980, the union government established the Department of Environment. It became the Ministry of Environment and Forests (MoEF) in 1985. This ministry initiates and oversees the implementation of environmental policies, plans, laws, and regulations.

Environmental Legislation in India

Following the Stockholm Conference, India began enacting environmental laws. Initially, the laws were not very different from the general body of law. For example, the Water Act of 1974 was very much like other laws and created another agency-administered licensing system to control effluent discharges into water.

The Bhopal gas tragedy (see Chapter 6) changed the situation. In the 1990s, a spate of laws were passed covering new areas like vehicular emissions, noise, hazardous waste, transportation of toxic chemicals, and environmental impact assessment.

Further, the old licensing regime was supplemented by regulatory techniques. The new laws included provisions like public hearings, citizens' right to information, deadlines for technology changes (in motor vehicles, for example), workers' participation, and penalties on higher management of companies. The powers of the enforcing agencies like the pollution control boards were also increased to levels much higher than before.

Enforcement of Environmental Laws in India

The implementation of environmental laws continues to be poor. The government agencies have vast powers to regulate industries and others who are potential polluters. They are, however, reluctant to use these powers to discipline the polluters.

The poor performance of the government agencies in enforcing the laws has compelled the courts to play a proactive role in the matter of environment. They have directly responded to the complaints of citizens or public interest litigation and assumed the roles of policymakers, educators, administrators, and, in general, friends of the environment. The environmental lawyer M.C. Mehta has played a remarkable role in this process.

Many petitions filed in courts by individual citizens, groups, and NGOs have led to major decisions given by the courts forcing the governments to act against pollution and degradation of the environment. The Supreme Court, for example, has in numerous cases issued directions to close down or shift factories, change to less-polluting technologies, implement environmental norms, and so on.

Box 12.2 Major Environmental Laws and Regulations of India

- Environment (Protection) Act of 1986: An enabling law that provides powers to the government for framing various rules and regulations.
- Environment (Protection) Rules: These rules set the standards for emission or discharge of environmental pollutants.
- Environmental Impact Assessment (EIA): Since 1994, EIA is mandatory for certain types of projects. The regulations require the project proponent to submit an EIA report and hold a public hearing on the project.
- Air (Prevention and Control of Pollution) Act of 1981: The objective of this Act is to provide for the prevention, control and abatement of air pollution.
- Water (Prevention and Control of Pollution) Act of 1974: The objectives are the prevention and control water pollution and the maintenance or restoration of the wholesomeness of water.
- Forest Conservation Act of 1980: It provides for the protection and conservation of forests.

In addition, there are laws concerning wildlife protection, hazardous waste management, coastal zone management, and so on.

As a result of this 'judicial activism', hundreds of factories have installed effluent treatment plants and there is a much greater environmental awareness among the bureaucracy, police, and municipal officials, all of whom are involved in implementing the court's orders. Judicial activism has certainly helped the cause of environment, but there is also the view that the courts should not be taking over the role of the executive.

We also have in India a large number of environmental NGOs, which work for conservation, organize campaigns, watch for violation of laws, and so on.

On the whole, the world's response to the environmental crisis has been muted. We have not done enough and we have not done things on time. Even as we dither, the circle is closing on us.

Box 12.3 The Story of M.C. Mehta: The Green Avenger

Delhi lawyer M.C. Mehta visited the Taj Mahal for the first time in early 1984. He noticed that the monument's marble had turned yellow and was pitted, as a result of pollutants from nearby industries. Shocked by what he saw, Mehta filed an environmental case in the Supreme Court of India on the issue.

In 1993, after a decade of hearings, the Supreme Court ordered 212 small factories surrounding the Taj Mahal to close because they had not installed pollution control devices. Another 300 factories were put on notice to do the same.

Mahesh Chandra Mehta did not stop with the Taj Case. In 1985, he filed a petition in the Supreme Court against the factories that were polluting the River Ganga. Many positive actions to prevent the pollution of the river resulted from this case.

Called the 'Green Avenger', Mehta has continued his crusade for a cleaner environment. At the last count, his cases numbered more than 50 in the Supreme Court alone. He has successfully fought cases against industries that generate hazardous waste. He succeeded in obtaining a court order to make lead-free gasoline available. He also worked to ban intensive shrimp farming and other damaging activities along India's coast.

Through his work, Mehta has set the national agenda in the fields of water and air pollution, vehicular emission control, conservation of the coastal zone, and the translocation of heavy industry away from urban areas. Almost single-handedly, he has obtained more than 40 landmark judgements and numerous orders from the Supreme Court against polluters. In addition, responding to his petition, the court ordered the inclusion of environmental education as a compulsory subject in all the schools, colleges, and universities of the country.

No other environmental lawyer in the world has perhaps done this much. He has inspired many lawyers in the lower courts to take up environmental cases. Mehta won the Goldman Environmental Prize in 1996 and the Ramon Magsaysay Award for Public Service in 1997.

He has set up the M.C. Mehta Environmental Foundation. It works for the protection of environment, rights of the people for clean and fresh water and air, promotion of sustainable development, and protection of cultural heritage of India.

Box 12.4 The Peace of Wild Things

When despair for the world grows in me
and I wake in the night at the least sound
in fear of what my life and my children's lives may be
I go and lie down where the wood drake
rests in his beauty on the water, and the great heron feeds
I come into the peace of wild things
who do not tax their lives with forethought
of grief. I come into the presence of still water.
And I feel above me the day-blind stars
Waiting with their light. For a time
I rest in the grace of the world, and am free.

—Wendell Berry*

*(b. 1934); American farmer, writer, and environmentalist.

END PIECE

The way humanity is hastening towards self-destruction reminds one of the Chinese proverb:

If we do not change the direction in which we are going, we are likely to end up precisely where we are headed.

EXPLORE FURTHER

Books

CSE, 2007, *The Anil Agarwal Reader (Vols. 1–3)*, New Delhi: Centre for Science and Environment.

Divan, Shyam and Armin Rosencranz, 2001, *Environmental Law and Policy in India*, second ed., New Delhi: Oxford University Press.

World Commission on Environment and Development, 1987, *Our Common Future*, Oxford: Oxford University Press.

Websites

Goldman Prize: www.goldmanprize.org
M.C. Mehta Environmental Foundation: www.mcmef.org

Film

Anil Agarwal: www.youtube.com/watch?v=SNjdQPV-z9g&feature=related

13 HOPE FOR THE PLANET AND HOW YOU CAN HELP

It often happens that I wake at night and begin to think
about a serious problem and decide I must tell the Pope about it.
Then I wake up completely and remember that I am the Pope.

—POPE JOHN XXIII*

THE STORY OF CHIPKO: HUG THE TREES, STOP THE LOGGING

On 26 March 1974, a group of men arrived stealthily in the forest next to Reni village in the Garhwal district of Himalayas. They had been sent by a contractor to begin cutting down 2,500 trees in the forest. The contractor had made sure that all the men were away on that day.

Word of the arrival of the axe-men spread in the village and the women came out of their houses. About 25 of them, led by Gaura Devi, confronted the contractor's men. They pleaded with the men not to start the felling operations, but the men responded with threats and abuses. As the confrontation continued, more women joined the protest. Ultimately, the men were forced to leave, since the women did not budge.

This small event was a milestone in the Chipko Movement, which became known all over the world as a symbol of people's action in preventing the destruction of the environment. '*Chipko*' means 'to cling' or 'to hug hard'. The

*(1881–1963); born Angelo Giuseppe Roncalli, 261st Pope of the Catholic Church.

vision of women hugging the trees and daring the axe-men to cut them before cutting the trees fired the spirit of environmentalists the world over. It remains one of the well-known environmental movements of the world.

It is not clear when and where the movement to save the trees got the name of Chipko. However, the movement spread rapidly across the Himalayan region in the 1970s. It was led by dedicated activists like Chandi Prasad Bhatt and Sunderlal Bahuguna.

The contract system allowed rich contractors from the plains to make huge profits from cutting trees on the hills. The Chipko movement was a response of the hill communities to the unfair and destructive nature of this system. As Raturi, the folk-poet of the movement put it,

Embrace the trees in the forests
And save them from being felled!
Save the treasure of the mountains
From being looted away from us!

There was also the catchy slogan of the movement:

What do the forests bear?
Soil, water and pure air!

Chipko forced the government to abolish the contract system and ban logging in the area for 15 years.

The Chipko Movement brought unprecedented energy and direction to environmental preservation in India. It spawned action in other places: Appiko Movement in Karnataka, Chilka Lake Agitation in Orissa, and Narmada Bachao Andolan.

Yet, Chipko lost its steam before its immense potential could be realized. Exaggeration of its strengths, excessive adulation, ego clashes among the key players—all contributed to the sapping of its energy. However, we can surely hope that new Chipkos will arise in the future.

WHAT CAN WE LEARN FROM THE STORY OF THE CHIPKO MOVEMENT?

Above all, the story of Chipko gives us hope that when people get together for a common cause, unexpected changes can occur. It shows also how a simple but powerful idea can spread spontaneously and rapidly in society. That gives us hope.

FROM DESPAIR TO HOPE

This (last) chapter of the book is addressed directly to you, the reader.

The book has been a catalogue of environmental destruction, sprinkled with signs of positive change. The state of the world can easily drive you to despair. But how can you move from a sense of despair to one of hope?

The first step is to realize that you have the power to change things. Mahatma Gandhi said:

> *The difference between what we do and what we are capable of doing would suffice to solve most of the world's problems.*

There are three levels at which you can exercise the power you have:

- Power of One: You can take right actions as an individual.
- Power of a Group: You can join (or create) a group or community of like-minded people working for positive changes.
- Power of a Movement: You can join any one of the social and environmental movements that are determined to change the world.

WHO IS THE ENEMY?

You will be led to act, when you ask: Who really is responsible for the environmental crisis? Who is the real enemy?

- Rich societies like that of the US that consume and pollute a lot?
- Governments and political parties that are only concerned about the next elections and hence do not take the necessary tough decisions?
- Companies that use natural resources and just throw away the waste?
- Corrupt officials who let the companies pollute?
- Municipalities that do not clear the garbage?
- Your neighbour, who throws the garbage on the street?

It is easy to blame others for the crisis, though some of them may well deserve a share of the blame. However, a little reflection shows that, at the very basic level, you are the cause and each one of us is the cause. As the cartoon character Pogo puts it, 'We have seen the enemy and it is us!'

The root causes of the crisis include:

- Your consumption and desire for more.
- The waste that you create and throw away mindlessly.
- Your NIMBY (Not In My Back Yard) syndrome: The practice of objecting to something that would affect you or take place in your locality, even though it would benefit many others.
- The corruption that you support, directly or indirectly.

- Perhaps also the unsustainable or polluting nature of your work or your organization.

Your actions such as these lead to increasing production of goods and services, overexploitation of natural resources, more waste and pollution, increased corruption, and so on.

If you agree that you are part of the problem, one could ask you, 'Can you become a part of the solution?' Your response could well be, 'What can I do? I am just a powerless individual. How can I fight against the bureaucrats, politicians, or industrialists?'

Power of One

Many individuals all over the world have indeed tried to become part of the solution. They have acted to save the environment and brought about real changes. There are many such inspiring stories, a few of which were given (some briefly) in the previous chapters:

- Water conservation by Rajendra Singh, Anna Hazare, and Vilas Salunke.
- The campaign for the cleanup of Love Canal and other toxic dumps by Lois Gibbs (see Chapter 6).
- Saving or planting of trees by Thimmakka, Wangari Maathai, and Julia Butterfly Hill.
- Planning and building Curitiba, an environment friendly city, by Jaime Lerner.
- Restoring Surat after the plague by S.R. Rao.
- Advocacy work of Anil Agarwal (see Chapter 12).
- Public interest litigation by M.C. Mehta.

There was also the fictional, but inspiring story, 'The Man Who Planted Trees' (Prologue to this Part).

Box 13.1 gives brief accounts of some more inspiring individuals.

Many more such environmental heroes can be found across the world. True tales of this kind have many uses:

- They can inspire us to act in our lives and in our context.
- By presenting environmental issues in the context of real places and events, they are more effective in spurring action than mere fear-inducing information.
- Through the stories, we understand the differences and commonalities of environmental problems and solutions around the world.
- The tales can be a great tool in education and creation of awareness among children and the youth.

Reading these extraordinary stories, we may be tempted to ask, 'What makes ordinary individuals take up seemingly impossible tasks in the face of heavy odds?' Clearly, given the depressing data, the driving force could not be just

Box 13.1 *Inspiring Individuals*

- Abdul Karim: Working alone and without any funds for over 25 years, Abdul Karim has created a forest on a barren hillside in Kerala.
- Chico Mendes: As a rubber tapper in the Brazilian Amazon, Chico Mendes fought for the conservation of forests. He met with some success, but was killed by a powerful local rancher, who was converting the forests into grazing land. Chico's life has been an inspiration for others.
- Ray Anderson: As the owner of Interface, a carpet-making company, Ray realized one day that he was 'a plunderer of nature'. He has set out to make his company a totally sustainable green corporation. As he reduced energy use and waste creation, the profits actually went up. He is a great model for other businessmen.
- Romulus Whitaker: He set up a snake park and a crocodile bank in Chennai and has been breeding endangered species of both snakes and crocodiles. He also created sustainable livelihoods for the Irulas, who are experts in catching snakes and extracting their venom.
- Baba Seechewal: A Sikh holy man, Baba Balbir Singh Seechewal, with his followers, has transformed the dirty Kali Bein, a holy river associated with Guru Nanak, into a clean water body by removing weeds, effluents, sewage, and other pollutants. This action raised the water tables in the area and restored farming. He has also demonstrated how treated sewage water could be used for organic farming. Now people are asking him to clean up other polluted rivers in Punjab.

optimism, which arises from the head. It can only be hope that springs from the heart, as described by Vaclav Havel, writer and first President of the Czech Republic:

Hope is an orientation of the spirit, an orientation of the heart; it transcends the world that is immediately experienced, and is anchored somewhere beyond its horizons.

Havel (who led resistance groups against the Soviet Union), Nelson Mandela (who led the struggle against apartheid in South Africa), Martin Luther King, Jr., (who led the Civil Rights Movement in the US) and others like them were kept going, not by a sense of optimism, but undying hope. That hope did triumph at the end. In the same way, the environment may also be saved from a disaster.

What You Can Do

You may not be ready yet to become an inspiring hero, but you can still do your bit in healing this planet. You could take three steps:

- Be informed: Read books and magazines and scour websites to find out for yourself where the world stands (See *Appendix A* for references).
- Examine yourself and your lifestyle by asking questions.
- Take action at home and work place, in the neighbourhood, in your city, and so on.

The previous chapters have given many suggestions for action. You can find some more in Box 13.3.

Power of a Group

Every city has many NGOs that take up local environmental issues and conduct campaigns during weekends. There is strength in numbers and you may see better results when you work with such groups. The famous anthropologist Margaret Mead said:

Never doubt that a small group of thoughtful committed citizens can change the world. Indeed, it is the only thing that ever has.

You can also go beyond just working with local environmental groups and move into a community or create one. Often, such a community grows from the vision of an individual or a small and dedicated group. Some have ecology as a focus, while others are oriented towards religion and spirituality. Some communities concentrate on social work. In India, there are also Gandhian communities, inspired by the ideals of the Mahatma. Box 13.4 gives five examples of communities that focus (among other things) on the environment—three in India and two in other countries. There are many more, large and small as well as old and new.

CHIPKO
MOVEMENT

Box 13.2 Some Questions to Ask of Oneself

- How much water and energy do I use every day?
- What do I buy? From where?
- What kind of food do I eat?
- What do I throw away? Where does the waste go?
- How do I travel?
- What types of medicines do I take?
- How do I build my house?
- How do I treat other living beings, plants?

Box 13.3 What More Can You Do to Save the Environment

- Try to reduce your ecological footprint by examining the source of everything that you consume.
- Find a job in which you can work from home on all or most days.
- If there are ponds, lakes, or wetlands in or near your area, join groups trying to save them. If there is no group and if a water body is in peril, organize a group to make representations to the government and to create awareness in the community.
- Organize a waste management system in cooperation with ragpickers: Residents keep organic waste and recyclable waste separately. Ragpickers collect the two kinds of waste at the doorstep and take them to a central location. They compost the organic waste and sell the manure. They also sort the recyclable waste and sell it to the traders. Such arrangements are working in many Indian cities.
- Cooperate with one another to keep the streets and the neighbourhood clean and to plant trees.
- Celebrate festivals like Diwali together, thereby minimizing toxic fumes and noise.
- Acquaint yourself with the basic environmental laws and take action when you come across any violation. File complaints as a group with the appropriate authorities.

Power of a Movement

If the planet is to be saved from destruction, we need many Chipko-type people's movements. Paul Hawken, who has been studying such movements, believes that there are up to two million organizations in the world working toward ecological sustainability, social justice, and indigenous rights: 'It is a global humanitarian movement arising from the bottom up. It is in fact a coherent, organic, self-organized, worldwide movement involving tens of millions of people dedicated to positive change.'

Hawken puts it this way:

> *If you look at the science that describes what is happening on earth today and aren't pessimistic, you don't have the correct data. If you meet the people in this unnamed movement and aren't optimistic, you haven't got a heart.*

Joanna Macy, American buddhist scholar and peace activist, is sure that a 'Great Turning' is occurring in the world: 'This turning is a shift from the industrial growth society to a life-sustaining civilization. Future generations will look back at the epochal transition we are making to a life-sustaining society. It is happening now. To see this as the larger context of our lives clears our vision and summons our courage.'

David Korten, Macy's colleague and author of the book *The Great Turning*, believes that everything is going to change. He suggests that we should embrace the crisis as an opportunity and 'move from Empire to Earth Community, from Dominator story to Partnership story, and from Competition to Cooperation' (Box 13.5). Alex Steffen, author of *Worldchanging*, is clear that 'real solutions already exist for building the future we want. It is just a matter of grabbing hold and getting moving.'

WHAT IS STOPPING US FROM ACTING TO SAVE THE PLANET?

Taking actions suggested in this book is necessary, but not sufficient. A profound change is possible only if all of us change our mindsets. Albert Einstein said, 'You cannot solve a problem with the same mindset which created it in the first place.'

We mistakenly see a difference between other beings and ourselves, between nature and ourselves. When we truly begin to see ourselves as an indivisible part of the universe, there will be a fundamental shift in our mindset.

Perhaps the first step in saving the planet is to fall in love with its beauty. The second step is to hear within us the sounds of the earth crying. The third step is to put this new understanding into action.

Good luck and welcome to a state of Blessed Unrest!

Box 13.4 Examples of Communities

- Gaviotas: It was established in 1971 by Paulo Lugari on 25,000 acres of land on the impoverished eastern plains of Colombia. The Gaviotans are about 200 researchers, students, and workers and they produce 70 per cent of food and energy needed. The place is run by consensus and unwritten rules. The residents get good wages, free meals, medical care, schools, and housing. Over the years, Gaviotans have come up with an extraordinary number of inventions including efficient wind turbines, solar devices, self-cooling rooftops, cooling wind corridors, corkscrew manual well digger, and so on. Since Gaviotas does not patent its inventions, its ideas have spread in the region, from Central America to Chile.
- Findhorn: This community in Scotland came up in 1962 as an answer to the question: How to transform our human settlements into full-featured sustainable communities, harmoniously and harmlessly integrated into the natural environment? It is one of the largest holistic communities in the world. Apart from conducting regular workshops, it also promotes the Global Ecovillage Network (GEN). Ecovillages stand for sustainability not only in environmental but also in social, economic, and spiritual terms.
- Auroville: Called also the 'City of Dawn', this is an international township in Tamil Nadu, near Puducherry, dedicated to human unity. It was founded in 1968 by the French visionary known as 'The Mother', a spiritual collaborator of the sage Sri Aurobindo. It is a remarkable story of ecological transformation achieved by a group of people (now about 2,000 in number) from 35 different countries including India. They have planted three million trees to convert barren land into lush greenery. They have also experimented with a range of environment friendly technologies covering water, energy, waste, building, and farming. Local biodiversity and medicinal knowledge is being recorded and revived.
- Timbaktu Collective: This community was established in 1990 on 32 acres of dry, degraded land in the drought prone Anantapur district of Andhra Pradesh. Slowly over the years

cont'd ...

... cont'd

not only Timbaktu but also the surrounding hills have greened themselves while insects, birds, and animals have reappeared. A small community of volunteers, committed to developmental and ecological regeneration, has settled here. Timbaktu's vision is to stop the degradation of the land in the area and to find ways to reverse it. The aim is also to develop alternative lifestyles that are sustainable and provide more liberty and happiness, than those based upon exploitation.

- Navadarshanam: This small organization is located on 115 acres of land in Tamil Nadu, 50 km. south of Bangalore. It investigates ecological and spiritual alternatives to the modern way of living and thinking. The areas of focus include eco-restoration of land, renewable energy, ecological architecture, healthier food and the science-spirituality connection. Navadarshanam organizes workshops and meetings on these aspects.

Box 13.5 Change the Story, Change the Future

Every society essentially plays out a story. One story has the theme, 'Man is an integral part of nature and is not separate from it.' This was the story of many indigenous peoples of the world. Today the dominant story tells us again and again in every way:

- Produce more and more products and services, be a strong economy, and a powerful nation.
- Make more money, be happier.
- Consume more, enjoy more, that is your right.
- Compete or you will perish.

We can change the story to a new one that tells us:

- We are the stewards of the earth.
- Degrading the earth with the waste products of our luxurious living is a sin.
- The right path is to live as frugally and simply as possible.
- Reducing consumption is not a sacrifice; it is an elevating experience.
- Partnership and cooperation are the keys to survival.

Box 13.6 Conclusion

The spirit of human solidarity and kinship with all life is strengthened
when we live with reverence for the mystery of being,
gratitude for the gift of life,
and humility regarding the human place in nature.

—FROM 'THE EARTH CHARTER' (2000)

If the Earth's environment and humanity have to survive, we cannot continue to exploit nature the way we have done over the past 100 years or more.

The people in the rich countries cannot continue with their wasteful, patently unsustainable, consumption and those in the poor countries cannot continue to yearn for such lifestyles.

Humankind has to move towards a simpler life in harmony with nature.

Box 13.7

Here is a moving excerpt from a poem by the American poet, essayist, and feminist, Adrienne Rich (b. 1929). It occurs towards the end of the poem 'Natural Resources' in her book *Dreams of a Common Language*.

My heart is moved by all I cannot save:
so much has been destroyed

I have to cast my lot with those
who age after age, perversely,

with no extraordinary power,
reconstitute the world.

A passion to make, and make again
where such un-making reigns.

END PIECE

Joining social movements and being part of a sustainable future is exhilarating, but not easy. It is akin to any creative work, about which the American dancer Martha Graham said:

[There is] no satisfaction whatever at any time. There is only a queer, divine dissatisfaction, a blessed unrest that keeps us marching and makes us more alive than others.

EXPLORE FURTHER

Books

Dogra, Bharat, n.d., *When Villagers Hugged Trees to Save Them: An Inspiring Story from Himalayan Villages in India*, New Delhi: Self-published.

Hawken, Paul, 2008, *Blessed Unrest: How the Largest Movement In the World Came Into Being and Why No One Saw it Coming*, New York: Penguin Books.

Korten, David, 2007, *The Great Turning: From Empire to Earth Community*, San Francisco: Berrett-Koehler Publishers.

Shiva, Vandana and Jayanto Bandhyopadhyay, 1986, *Chipko: India's Civilisational Response to the Forest Crisis*, New Delhi: INTACH.

Steffen, Alex (ed.), 2006, *Worldchanging: A User's Guide for the 21st Century*, New York: Harry N. Abrams.

Websites

www.ecotippingpoints.org
www.worldchanging.com

APPENDIX A
Explore Further

BOOKS

- Bruges, James, 2001, *The Little Earth Book*, second edn., Bristol: Alastair Sawday Publishing.
- Caldicott, Helen, 2009, *If You Love This Planet: A Plan to Save the Earth*, New York: W.W. Norton.
- Callenbach, Ernest, 1999, *Ecology; A Pocket Guide*, Indian edn., Hyderabad: Universities Press.
- Meadows, Donella H., J. Randers, and D. L. Meadows, 2004, *Limits to Growth: The 30-Year Update*, White River Junction, VT, USA: Chelsea Green.
- Meadows, Donella H., D. L. Meadows, J. Randers, and W.W. Behrens, 1972, *The Limits to Growth*, New York: Universe Books.
- Raghunathan, Meena and Mamata Pandya (eds), 1999, *The Green Reader: An Introduction to Environmental Concerns and Issues*, Ahmedabad: Centre for Environmental Education.
- Rajagopalan, R., 2011, *Environmental Studies: From Crisis to Cure*, second edn., New Delhi: Oxford University Press.

WEBSITES

United Nations

- Convention on Biological Diversity (UNCBD): www.biodiv.org
- Equator Initiative (Projects that alleviate poverty through biodiversity conservation): www.undp.org/equatorinitiative

- Food and Agriculture Organization (FAO): www.fao.org
- Global Environment Facility (GEF): www.gefweb.org
- Intergovernmental Panel on Climate Change (IPCC): www.ipcc.ch
- UNEP-World Conservation Monitoring Centre (UPEP-WCMC): www.unep-wcmc.org
- United Nations Environment Programme (UNEP): www.unep.org
- World Health Organization (WHO): www.who.int
- World Meteorological Organization (WMO): www.wmo.ch

Biodiversity issues

- Biodiversity Hotspots: www.biodiversityhotspots.org
- Biodiversity: www.iisd.org
- Global Biodiversity Information Facility (GBIF): www.gbif.net
- Global Biodiversity Outlook: www.biodiv.org/gbo/gbo-pdf.asp
- Plants for a Future: www.pfaf.org
- Red List of Threatened Species: www.redlist.org

Water

- Water Supply and Sanitation Collaborative Council (WSSCC): www.wsscc.org
- World Water Council (WWC): www.worldwatercouncil.org
- World Water Forum: www.worldwaterforum.org

Energy

- Alliance to Save Energy: www.ase.org
- Centre for Renewable Energy and Sustainable Technology (CREST): www.solstice.crest.org
- Energy and fossil fuel crisis: www.dieoff.org
- Energy Conservation Center of Japan (ECCJ): www.eccj.or.jp
- International Solar Energy Society (ISES): www.ises.org
- Oil Depletion: www.lifeaftertheoilcrash.net

Climate change

- Biodiversity and climate change: www.wri.org/ffi/climate/kyotfrst.htm
- Carbon sinks: www.weathervane.rff.org/features/feature050.html
- Climate Action Network (CAN): www.climatenetwork.org
- Climate change: www.changingclimate.org
- Climate change: www.pewclimate.org
- Forest Carbon and Local Livelihoods: www.cifor.cgiar.org/
- Greenhouse Gases: www.ghgonline.org

Information portals

- Environmental Information Service (ENVIS) India: www.envis.nic.in
- Environmental News Network (ENN): www.enn.com
- International Environmental Law Research Centre (IELRC): www.ielrc.org
- Online books on farming, etc.: www.journeytoforever.org
- Non-profit research tool: www.mindfully.org
- Soil and Health Library: www.soilandhealth.org

PERIODICALS

Indian environmental periodicals

- *Down To Earth*: A fortnightly published by the Centre for Science and Environment (CSE), New Delhi. Once a month it includes *Gobar Times,* a supplement for children. (www.cseindia.org)
- *Hindu Survey of Environment:* Published annually by the Hindu Group, Chennai. (www.thehindu.com/publications/)
- *Sanctuary Asia* and *Sanctuary Cub* (for children): Published by Bittu Sahgal. (www.sanctuaryasia.com)
- *Terra Green:* Published monthly by The Energy and Resources Institute (TERI), New Delhi. (http://terragreen.teriin.org/)

Indian periodicals that carry articles on environment

- *Agenda* (www.infochangeindia.org)
- *Civil Society* (www.civilsocietyonline.com)
- *Frontline* (www.frontlineonnet.com/)
- *Seminar—The Monthly Symposium* (www.seminar-india.com)

Selected foreign environmental periodicals

- *E—The Environmental Magazine* (www.emagazine.com)
- *Earth Island Journal* (www.earthisland.org)
- *Green Teacher* (www.greenteacher.com)
- *Mother Earth News* (www.motherearthnews.com)
- *OnEarth* (www.onearth.org)
- *Orion* (www.orionmagazine.org)
- *Resurgence* (www.resurgence.org)
- *Sierra* (www.sierraclub.org)
- *Solutions* (www.thesolutionsjournal.com)
- *The Ecologist* (www.theecologist.org)

APPENDIX B

Selected List of Environmental Organizations/ Departments

GOVERNMENT OF INDIA

- Central Pollution Control Board (CPCB): www.cpcb.nic.in
- Indian Renewable Energy Development Agency (IREDA): www.ireda.in
- Ministry of Environment and Forests (MoEF): http://envfor.nic.in
- Ministry of New and Renewable Energy: http://mnes.nic.in
- Ministry of Water Resources: http://wrmin.nic.in
- National Biodiversity Authority: www.nbaindia.org
- National Environmental Engineering Research Institute (NEERI): www.neeri.res.in
- National Institute of Oceanography (NIO): www.nio.org
- National River Conservation Directorate (NRCD): http://envfor.nic.in/nrcd/
- Wildlife Institute of India, Dehra Dun (WII): www.wii.gov.in

INDIAN ORGANIZATIONS AND INSTITUTIONS

- Ashoka Trust for Research in Ecology & the Environment (ATREE), Bangalore: www.atree.org
- Bombay Environmental Action Group (BEAG), Mumbai: www.beag.net
- Bombay Natural History Society (BNHS), Mumbai: www.bnhs.org
- Centre for Ecological Sciences (CES) at the Indian Institute of Science (IISc), Bangalore: http://wgbis.ces.iisc.ernet.in
- Centre for Environmental Education (CEE), Ahmedabad: www.ceeindia.org
- Centre for Indian Knowledge Systems (CIKS), Chennai: www.ciks.org
- Centre for Science and Environment (CSE), New Delhi: www.cseindia.org
- Environment Support Group (ESG), Bangalore: www.esgindia.org

- Exnora International, Chennai: www.exnora.org
- Foundation for Ecological Security (FES), Anand: www.fes.org.in
- Foundation for the Revitalisation of Local Health Traditions (FRLHT), Bangalore: www.frlht.org
- Goa Foundation, Mapusa, Goa: www.goafoundation.org
- Greenpeace India, Bangalore: www.greenpeace.org/india/
- Himalayan Eco Horticulture Society (Ecoharts), Himachal Pradesh: www.ecohorts.org
- Indian Green Building Council (IGBC), Hyderabad: www.igbc.in
- Indian Youth Climate Network (IYCN): www.iycn.in
- International Campaign for Justice in Bhopal, Bhopal: www.bhopal.net
- Kalpavriksh Environment Action Group, Pune: www.kalpavriksh.org
- Ladakh Ecological Development Group (LEDG), Ladakh: www.ledeg.org
- M.S. Swaminathan Research Foundation, Chennai: www.mssrf.org
- Madras Crocodile Bank Trust, Chennai: www.madrascrocodilebank.org
- Navdanya, New Delhi: www.navdanya.org
- Nilgiri Wildlife and Environment Association (NWEA), Udhagamandalam: www.nwea.in
- People for the Ethical Treatment of Animals (PETA), Mumbai: www.petaindia.com
- Prayogpariwar, Maharashtra: www.prayogpariwar.net
- Salim Ali Centre for Ornithology and Natural History (SACON), Tamil Nadu: www.sacon.org
- Sankat Mochan Foundation, Varanasi: www.sankatmochanfoundationonline.org
- TERI (The Energy and Resources Institute), New Delhi: www.teriin.org
- Toxics Link, New Delhi: www.toxicslink.org
- Wildlife Protection Society of India (WPSI), New Delhi: www.wpsi-india.org/
- Wildlife Trust of India (WTI), Noida, Uttar Pradesh: www.wildlifetrustofindia.org
- WWF-India, New Delhi: www.wwfindia.org

SELECTED INTERNATIONAL ORGANIZATIONS AND INSTITUTIONS

- Audubon Society: www.audubon.org
- Earth Policy Institute (EPI): www.earth-policy.org
- Greenpeace: www.greenpeace.org
- International Union for Conservation of Nature (IUCN): www.iucn.org
- World Wide Fund for Nature (WWF): www.panda.org
- Worldwatch Institute: www.worldwatch.org
- Wuppertal Institute: www.wupperinst.org

SOURCES FOR FILMS

- Centre for Media Studies (CMS), New Delhi: www.cmsindia.org
- Centre for Science and Environment (CSE), New Delhi: www.cseindia.org
- TERI (The Energy and Resources Institute): www.teriin.org

APPENDIX C
Find out and be surprised! Question, Answers, Explanations

CHAPTER 1 PLANET IN PERIL: THREE STORIES

The Himalayan glaciers supply water to a large number of people in the region. Can you guess what percentage of the world population does this cover?

Answer: The Himalayan glaciers are estimated to supply water to 20–40 per cent of the world population. This is because the Himalayas are the source of the Indus Basin, the Yangtze Basin, and the Ganga-Brahmaptura, which are three of the world's primary river systems. Often, we forget that there is a Chinese (Tibetan) side to the Himalayas! In fact, the Himalayas straddle six countries: Bhutan, China, India, Nepal, Pakistan, and Afghanistan.

CHAPTER 2 THE GLOBAL ENVIRONMENTAL CRISIS

In a city on the Ganga, the coliform bacterial count is at least 3,000 times higher than the standard established as safe by the UN World Health Organization (WHO). Which city could it be?

Answer: Varanasi.

CHAPTER 3 WHY ARE WE IN A CRISIS?

According to a 2007 report of the World Wide Fund for Nature, the US had the second highest per capita ecological footprint in the world. Which country could have been the first in the list?

Answer: United Arab Emirates (UAE).

Dubai and other places in the UAE are located in a desert and outside temperature could be 120°F. They consume enormous amounts of energy because of desalinization plants, air-conditioned cars, skyscrapers, megamalls, chilled swimming pools, indoor ski slopes, artificial lakes, golf courses, and so on.

CHAPTER 4 A THIRSTY WORLD: SCARCITY OF WATER

How much water can one drop of oil contaminate, making the water undrinkable?

Answer: One drop of oil can make up to 25 litres of water undrinkable.

CHAPTER 5 A WARMING WORLD: ENERGY, GLOBAL WARMING, AND CLIMATE CHANGE

1. Until 1958, about 115 billion barrels of oil had been drilled out of the earth. Can you guess how long it took to extract the next 115 billion barrels of oil?

   Answer: The first 115 billion barrels were taken out in 100 years (1859–1958), while the next 115 billion barrels were extracted in just 10 years (1959–69)!

2. If a Chinese citizen consumed oil in amounts equal to an average American citizen, how much oil would China need every day?

   Answer: China would then need 90 million barrels of oil per day, which is 11 million more barrels than the world produced in one day in 2001.

CHAPTER 6 A TOXIC WORLD: WASTE, POLLUTION, AND ENVIRONMENTAL HEALTH

1. India manufactures over 4 million tons of plastics packaging. Can you guess how much of it ends up as waste soon after usage?

   Answer: Of the over 4 million tons of plastics packaging that is manufactured, over 50 per cent ends up as waste soon after usage.

2. In which city will you find Asia's largest market for trade in plastic waste?

   Answer: Delhi.

3. How much money does New York City spend every day for just carrying away its garbage?

   Answer: US$ 1 million a day.

CHAPTER 7 A WORLD WITHOUT LIFE: VANISHING FORESTS AND ENDANGERED SPECIES

How long has the cockroach as a species existed on earth?

Answer: 350 million years.

What is the secret of their survival? First, they can eat almost anything including dead insects, fingernail clippings, electrical wires, glue, paper, and soap! Second, they can live and breed almost anywhere except in the polar regions. Some of the cockroach species can survive for months without food or live a whole month on a drop of water.

How do they escape from their enemies? They have antennae that can detect minute movements of air. There are vibration sensors in their knee joints. They can move very fast and some even have wings.

Cockroaches can sample the food before it enters their mouths. Hence they are able to avoid any poisonous stuff. They keep themselves clean and eat their own dead.

A single Asian cockroach and its offspring can produce about 10 million new cockroaches in one year. Because of this speed of reproduction, they can quickly develop resistance to pesticides.

CHAPTER 8 A DYING OCEAN: POLLUTED SEAS AND COASTS

1. What percentage of all life on earth is found in the ocean?

   Answer: An estimated 50–80 per cent of all life on earth is found under the ocean surface and the oceans contain 99 per cent of the living space on the planet.

2. How much of the ocean space has been explored by us?

   Answer: Less than 10 per cent of the ocean space has been explored by humans. Almost 85 per cent of the area and 90 per cent of its volume constitute the dark, cold environment we call the deep sea.

CHAPTER 9 A WORLD OF PEOPLE: GROWING POPULATION AND URBANIZATION

1. What percentage of the world's population lives in the Northern Hemisphere?

   Answer: 90 per cent.
   The northern hemisphere contains most of the earth's land area.

2. During the twentieth century, the demand for one human need increased faster than the population growth. Which one?

   Answer: Water.
   During the twentieth century, global demand for water rose twice as fast as population growth.

CHAPTER 10 A HUNGRY WORLD: LAND, SOIL, FOOD, AND AGRICULTURE

1. There are more than one billion hungry people in the world. What is the number of obese people and why are they obese?

   Answer: About one million people in the world are hungry and an equal number are overweight. That includes a large number of poor people, who are forced to choose cheap calories leading to obesity.

2. What percentage of the world's cereal production is used as animal feed?

   Answer: About 35 per cent to 40 per cent of the world's cereal harvest and over 90 per cent of the world's soya is used for animal feed.

APPENDIX D
Glossary

Acid rain: Rain, mist, or snow formed when atmospheric water droplets combine with a range of man-made chemical air pollutants.
Aquaculture: The artificial production of fish in ponds or underwater cages.
Aquifer: An underground layer of rock or sand that contains water.
Asbestos: A fibrous silicate mineral used as construction material and insulation. It is very dangerous to health when the fibres are inhaled.

Background extinction: The gradual disappearance of species due to changes in local environmental conditions.
Biodegradable waste: Any waste item that breaks down into the raw materials of nature and becomes part of the environment in reasonable time.
Biodiversity: The numbers, variety, and variability of living organisms and ecosystems. It covers diversity within species, between species, as well as the variation among ecosystems. It is concerned also with their complex ecological interrelationships.
Biofertilizer: Living microorganisms, cultured and multiplied for use as fertilizer.
Biofuel: Fuel oil from the seeds of certain trees; it can be mixed with diesel and used in engines.
Biogas: Gas generated from human and animal waste.
Biological extinction: The complete disappearance of a species. It is an irreversible loss with not a single member of the extinct species being found on earth.
Biological pest control: The intentional introduction of predators, diseases, or parasites to control pests.

Biomedical waste: Waste that originates mainly from hospitals and clinics and includes blood, diseased organs, poisonous medicines, and so on.

Biopesticide: Pesticides derived from animals, plants, bacteria, and certain minerals.

Biotechnology: Technology that manipulates living organisms or cells to create a product or an effect.

Bottom trawling: The practice of fishing by scraping the sea floor with a net.

Carbon cycle: Cyclic movement of carbon in various forms from the environment to organisms and back to the environment.

CFCs (or Chlorofluorocarbons): Chemical compounds that were used worldwide as ideal refrigerants, until they were found to be depleting the ozone layer of the atmosphere.

Checkdam or *johad*: A small structure of earth and stones that blocks the path of any flow of water and helps recharge the groundwater.

CITES: United Nations Convention on International Trade in Endangered Animal and Plant Species that bans the hunting, capturing, and selling of threatened species.

Climate change: Changes in the earth's climate, especially those produced by global warming.

Coastal zone: Area extending from the high tide mark on land to the edge of the continental shelf, where there is a sharp increase in the depth of water.

Compost: Sweet-smelling and dark-brown manure produced by microorganisms, when they break down organic matter like food waste, leaves, paper, and wood.

Composting: The process of converting organic waste into fertilizer.

Convention on Biological Diversity (CBD): International convention for the conservation and sustainable use of the world's biodiversity.

Coral reefs: Colourful protective crust of limestone formed by colonies of tiny organisms called polyps.

Crude birth rate: The number of live births per 1,000 people in a population in a given year.

Crude death rate: The number of deaths per 1,000 people in a population in a given year.

DDT (Dichloro-diphenol-trichloroethane): An insecticide that protects crops and human beings from insects. Being harmful to organisms, it is now banned in many countries.

Desertification: Land degradation in arid and semi-arid areas caused by human activities and climatic changes.

Dioxins: Chemical compounds that are formed when we burn waste, plastics, coal, or cigarettes. They are highly toxic to humans and animals.

Ecological architecture: Architecture that seeks to minimize the ecological footprint of the house, building, or complex that is being designed and constructed.

Ecological extinction: The state of a species when so few members are left that the species can no longer play its normal ecological role in the community.

Ecological footprint: It is the area of the earth needed to sustain the life of an entity indefinitely (a person, a city, or a country). Thus it measures how much of a burden an entity is on this planet.

Ecological sanitation (Ecosan): A sustainable closed-loop sanitation system that uses dry composting toilets.

Ecology: The science that studies the relationships between living things and their environment. It is often considered to be a part of biology.

Ecosystem (or ecological system): A set of species living in an area and interacting with their environment.

Ecosystem service: The ecological service provided by an ecosystem such as the maintenance of the biogeochemical cycles, modification of climate, waste removal and detoxification, and control of pests and diseases.

Endangered species: A species that is facing a very high risk of extinction in the wild in the near future.

Endosulfan: A pesticide used to protect many crops from pest attacks.

Environment: The natural world in which people, animals, and plants live.

Environmental conservation: All the ways in which we protect nature and reverse the damage caused to the natural environment. When we prevent the killing of wildlife, or we turn a barren land into a forest, we are conserving the environment.

Environmental degradation: This term refers to the damage caused to the natural environment. When we cut down forests, or we pollute a river by dumping waste into it, we are causing environmental degradation.

Environmental ethics: Moral principles that try to define our responsibility towards the environment.

Environmental health: Those aspects of human health that are determined by physical, chemical, biological, social, and psycho-social factors in the environment.

Environmental Impact Assessment (EIA): The study of the likely short-term and long-term impact of a new project on the natural and social environment.

Environmental refugee: A person who has been displaced due to environmental degradation or a development project.

E-waste: Waste from discarded computers, mobile phones, and other electronic items.

Exclusive Economic Zone (EEZ): The area of the ocean up to a distance of 200 nautical miles from a country's shoreline over which the country has the exclusive right to exploit the resources.

Exponential growth: The growth of a quantity with time in such a way that the curve is relatively flat in the beginning, but becomes steeper and steeper with time.

Fauna: All the animals living in an area.
Flora: All the plants of a particular area.
Fluorosis: An ailment caused by the excess intake of fluoride.
Food chain: A sequence of species, in which each is the food for the next in the chain.
Food web: An interconnected set of food chains.
Fossil fuel: Fuel like coal and oil, formed millions of years ago from dead plants.
Fuel cell: An electrochemical unit that burns hydrogen to produce electricity.

Genetic engineering: The manipulation of the genes in an organism to change its characteristics, for example, moving a favourable gene from one organism to another.
Global warming: Warming of the earth's atmosphere due to the increase in the concentrations of gases like carbon dioxide.
Green Revolution: The rapid increase in world food production, especially in the developing countries, during the second half of the twentieth century, primarily through the use of lab-engineered high-yielding varieties of seeds.
Greenhouse gas: A gas like carbon dioxide that surrounds the earth and prevents some of the sun's heat from being reflected back out again.
Groundwater: Water contained underneath the earth's surface.

Habitat: The area where a species of plant or animal is particularly adapted to live.
Hubbert Curve: A curve proposed by the geophysicist M. King Hubbert that describes the pattern of oil availability in a field over time.

Idea of Progress: The belief that humankind would move on an unceasing path of better material conditions and a better life through economic and industrial development by exploiting natural resources.

Kyoto Protocol: An international agreement to cut greenhouse gas emissions.

Landfill: An area on which the city's garbage is dumped.

Mangroves: A unique salt-tolerant tree that grows in the coastal zone.
Marine Protected Area (MPA): Area set up by a country to protect a marine ecosystem, its natural processes, habitats, and species.

Mass extinction: Permanent loss of large numbers of species over a relatively short period of geological time.

Methyl Isocyanate (MIC): Highly poisonous chemical used in pesticide manufacture.

Montreal Protocol: An international agreement to replace ozone-depleting substances with safer ones.

NIMBY: It stands for 'Not in my Backyard'. It refers to our attitude of worrying only about our home and surroundings.

Non-renewable energy source: An energy source that is limited in supply and gets depleted by use.

Organic farming: A method of farming that does not use chemical fertilizers and chemical pesticides. It is a return to the traditional methods like crop rotation, use of animal and green manures, and some forms of biological control of pests.

Ozone layer: A layer of ozone that exists in the upper atmosphere, or stratosphere, between 10 and 50 km. above the earth.

PCBs (Polychlorinated biphenyls): A group of toxic chemical compounds that are very stable, have good insulation qualities, are fire resistant, and have low electrical conductivity. PCBs can pass through food chains.

Photovoltaic cell: A device that converts solar energy directly into electricity.

Protected area: Area in which the biodiversity and wildlife are protected from human exploitation. National parks, sanctuaries, and biosphere reserves are examples.

Rainforest: A type of forest found in the hot and humid regions near the equator. These regions have abundant rainfall and little variation in temperature over the year.

Rainwater harvesting: Harvesting rainwater where it falls, either by collecting and storing it, or by letting it recharge the groundwater.

Recycle: To convert waste back into a useful form.

Reduce: To reduce the amount of things we buy and consume.

Renewable energy source: An energy source that is replenished by natural processes and hence can be used indefinitely.

Reuse: To extend the life of an item or to find a new use for it.

Sacred grove: A forest patch, considered sacred and protected by the local community.

Salinization of soil: The building up of salt in the soil following the evaporation of excess water.

Seventh Generation Principle: Being cautious while taking decisions in a community by considering the effects of each decision on the next seven generations.

Sludge: The material left behind after the treatment of sewage; it is often toxic.

Social forestry: The planting of trees, often with the involvement of local communities, on unused land and wasteland.

Species: A set of organisms that resemble one another in appearance and behaviour. The organisms in a species reproduce naturally among themselves.

Species extinction: The situation in which no members of the species are found to exist anywhere on earth.

Sustainable development: Development that meets the needs of the present without compromising the ability of future generations to meet their own needs.

Threatened species: A species that is facing a high risk of global extinction.

Topsoil: The topmost layer of the soil, where most plant roots, microorganisms, and other animal life are located.

Toxic waste: Waste that contains poisons, which could kill certain organisms.

UNCLOS: The United Nations Convention on the Law of the Sea.

Urbanization: The process by which more and more people live and work in cities rather than in the rural areas.

Vermicomposting: Composting organic waste more efficiently using earthworms.

Water cycle: A process that continually recycles and transports water among the atmosphere, land, and ocean.

Water harvesting: The process of catching the rain when and where it falls.

Wetland: Land surface covered or saturated with water for a part or whole of the year.

INDEX